Progress in IS

"PROGRESS in IS" encompasses the various areas of Information Systems in theory and practice, presenting cutting-edge advances in the field. It is aimed especially at researchers, doctoral students, and advanced practitioners. The series features both research monographs that make substantial contributions to our state of knowledge and handbooks and other edited volumes, in which a team of experts is organized by one or more leading authorities to write individual chapters on various aspects of the topic. "PROGRESS in IS" is edited by a global team of leading IS experts. The editorial board expressly welcomes new members to this group. Individual volumes in this series are supported by a minimum of two members of the editorial board, and a code of conduct mandatory for all members of the board ensures the quality and cutting-edge nature of the titles published under this series.

More information about this series at https://link.springer.com/bookseries/10440

Jennifer Hehn • Daniel Mendez •
Walter Brenner • Manfred Broy
Editors

Design Thinking for Software Engineering

Creating Human-oriented Software-intensive Products and Services

Springer

Editors
Jennifer Hehn
Institute for Digital Technology Management
Bern University of Applied Sciences
Bern, Switzerland

Daniel Mendez
Department of Software Engineering
Blekinge Institute of Technology
Karlskrona, Sweden

Walter Brenner
Institute of Information Management
University of St. Gallen
St. Gallen, Switzerland

Manfred Broy
Department of Informatics
Technical University of Munich
Garching b. München, Germany

ISSN 2196-8705 ISSN 2196-8713 (electronic)
Progress in IS
ISBN 978-3-030-90593-4 ISBN 978-3-030-90594-1 (eBook)
https://doi.org/10.1007/978-3-030-90594-1

This Springer imprint is published by the registered company Springer Nature Switzerland AG.
The registered company address is: Gewerbestrasse 11, 6330 Cham, Switzerland

Introduction and Overview

The digitization of the economy and society is progressing rapidly. The SARS-CoV-2 pandemic has triggered an additional "digitalization surge" in which, among other things, working from home has become the new normal. Digitization means that the importance of software is constantly increasing. There is already almost no challenge in private and business life for which a software solution does not already exist. Today, there is software that seems almost perfect from the user's point of view. Apple, for example, has created a whole range of software solutions and devices, such as iPhones or iPads, which can be used effortlessly by people of different ages and with different affinities to the world of software and hardware, not least thanks to the work of Steve Jobs, who recognized the importance of non-functional aspects such as usability and appearance very early on. But not all software solutions serve their purpose. Very often, one stands amazed in front of software that does not do what it actually promised to do, or that does what it promised to do, but the functionality is difficult to use due to hard-to-use interfaces. We are all confronted with faulty or hard-to-use software. These deficits are surprising, because the development of software is no longer a young discipline, but now looks back on a history of almost 60 years.

The development of human-centric software solutions is becoming increasingly important. By human-centered software we mean software that is made by people for people and is user-appropriate and has functionality that is needed by users. The goal of this edited volume is to contribute to the improvement of the software development process by strengthening human centricity. In recent years, design thinking has emerged as an approach that combines innovation and human centricity. This approach has emerged in the environment of leading engineering faculties at universities such as Stanford University in Palo Alto, the Massachusetts Institute of Technology in Boston in the USA, and Carnegie Mellon University in Pittsburgh and has spread worldwide in recent years. Design thinking was originally focused on the development of new customer-focused physical products or to improve existing physical products. In recent years, design thinking has increasingly become a collection of established methods and approaches for the development of software and software-intensive products. These are the focus of this book. We consider a software-intensive system as "any system where software contributes essential influences on the design, construction, deployment, and evolution of the system as a whole" (IEEE; p. 1). Such systems typically consist of software as well as of hardware.

In this book, we focus on the first step of the software development process, requirements engineering. We want to add design thinking to the "method and toolbox" of requirements engineering. This edited volume was produced in cooperation between researchers at the Technical University of Munich, its associated institute fortiss, the Bleking Institute of Technology in Karlskrona in Sweden, the Institute for Digital Technology Management at Bern University of Applied Sciences, and the University of St. Gallen. The Technical University of Munich, fortiss, and the Bleking Institute of Technology contributed their extensive knowledge and experience in software development and requirements engineering, while the University of St. Gallen and the Bern University of Applied Sciences brought its expertise in design thinking to the collaboration. The cooperation turned out to be not easy since design thinking and software engineering are two different worlds colliding. Requirements engineering always aims at the development of software and defines the basis for software development activities at an early stage. Design thinking is an open-ended method. Software can be the result but does not have to be. The focus is on satisfying user needs that have been found. Human centricity in design thinking is more than operating appropriateness. It is about meeting user needs. The personalities involved in design thinking and software engineering also have different characters. Design thinkers are often open personalities with very different educational backgrounds. Numerous design thinkers come from an artistic-creative environment or are trained designers. Many design thinkers are extroverted and communicative. Software engineers are very often people with a penchant for formal mathematical thinking. They have usually studied computer science and have a background in the natural sciences. Open-ended work is rather alien to them. They want to build correct software. Accordingly, it is difficult to bring the two cultures, worlds of thought, and personalities together. The collaboration was driven by the common goal of combining the best of both worlds.

On the one hand, we searched for the authors of the invited contributions through extensive research on the Internet, and on the other hand, we approached personalities from our personal environment and invited them to write a contribution. Our Internet research showed that there are only some researchers and people from practice who are already working on the combination of requirements engineering and design thinking. The existing "body of knowledge" in academia is still small and probably just emerging. The area we deal with in this edited volume is only at an early stage. With this book we want to give a starting signal for the development of a "body of knowledge." In the end, 12 authors or teams of authors accepted our invitation. One contribution comes from two of the editors. We would like to express our sincere thanks for these contributions.

The contributions in this book can be divided into three clusters: Cluster Method, Cluster Governance, and Cluster Use Cases. Figure 1 gives an overview.

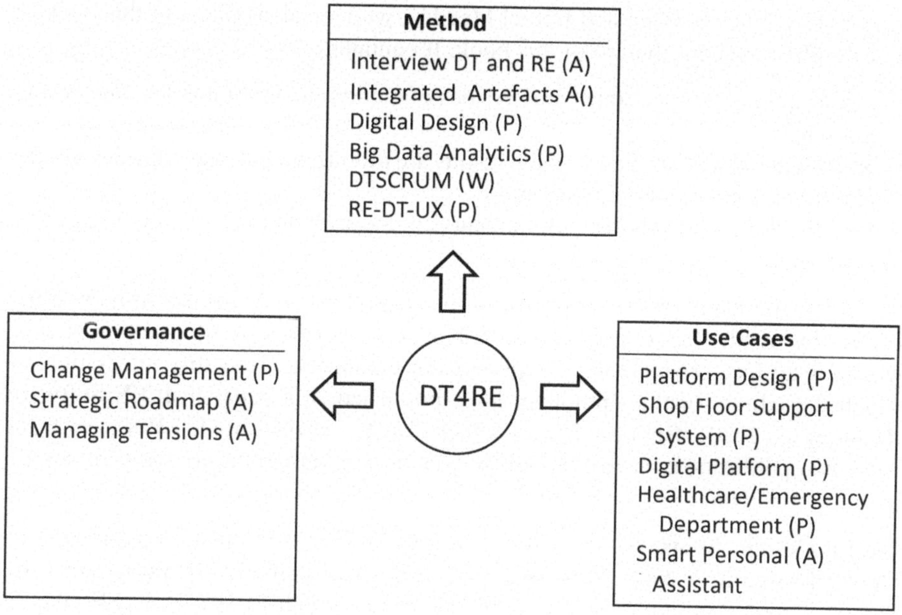

Fig. 1 Content structure of the twelve contributions in three clusters

Contributions from the field are marked with a P and those from academia with an A. Five papers deal with methodological issues, six papers present field reports, and three papers address governance issues.

We have organized the contributions in this book according to the three clusters we have identified. First come the contributions from the "Methodology" cluster, then the contributions from the "Governance" cluster, and finally come the contributions from the "Use Cases" cluster.

Cluster Method

Interview DT and RE The first contribution of this editor volume, an "Interview with Manfred Broy and Walter Brenner about design thinking and requirements engineering," is a fictitious conversation between two of the editors, Manfred Broy and Walter Brenner, as the two editors could not meet in person due to the pandemic. In this dialogue, they elaborate on the goals of this book and explain why combining requirements engineering and design thinking makes sense and can make a significant contribution to digitalization and business and society. The conversation also reveals that there are still many challenges to overcome before requirements engineering and design thinking are "merged."

Integrated Artifacts Combining design thinking and software requirements engineering to create human-centered software-intensive systems, the second paper is authored by Jennifer Hehn and Daniel Mendez, who are also editors of this book. It is really the core contribution of this book. It contains

- a description of the central characteristics of requirements engineering and design thinking
- an artifact model that shows which results are produced by design thinking and by classical requirements engineering
- how the "result worlds" of requirements engineering and design thinking can be connected

On the one hand, this contribution is based on Hehn's (2020) dissertation at the University of St. Gallen and is partly taken over verbatim as well as preliminary work on artifact orientation in requirements engineering, the first integration of which the editors of this anthology have published in a paper in IEEE Software (Hehn et al. 2020). Both publications focus on the artifact model. Passages from these two publications have been taken over in part verbatim into this core contribution of this book.

Digital Design In his chapter "From Design Thinking in Software Engineering to Digital Design as a New Profession - An Essay on Methods and Professions for Shaping Digital Solutions and Systems," Kim Lauenroth from Adesso SE takes a critical look at the use of design thinking in large software projects. Based on his experience in software engineering, Lauenroth is of the opinion that design thinking is not suitable as a general approach in large software projects. However, he concludes that design thinking methods are very suitable for designing innovative business processes. At the end of his chapter, he calls for a new discipline, digital design, which combines design thinking and software engineering. He succinctly describes that the computer science faculties, which are responsible for training in software engineering, have so far attracted only a few students who are interested in design. Lauenroth even goes so far as to draw up a competence profile for digital designers.

Big Data Analytics Michael Lewrick, author of several books on design thinking and Head of Innovation Labs at Deloitte Switzerland, shows in his chapter "The Hybrid Model: Combination of Big Data Analytics and Design Thinking" how design thinking can be combined with data analytics. Design thinking as a mixture of rather "soft" non-formal methods is combined with statistical analyses. Through this connection, it is possible very quickly to put measured data alongside the results of design thinking, which are of a more descriptive nature. The "gut feeling" of design thinking is supplemented, verified, and sometimes even discarded by data analysis. Lewrick refers to this as a "hybrid model" that combines design thinking and data science. In our view, the integration of elements from the fields of classical requirements engineering, design thinking, and data science is an essential component of future software engineering.

DTScrum Daniel Gadner from Internetstores and Michael Felderer from the University of Innsbruck deal with the connection of agile software development and SCRUM in their contribution "The Collective Process Framework DTScrum for integrating design thinking into Scrum." The focus of their contribution is the collective process framework. It represents a process model that combines agile software development and design thinking. The first step of their process model is the Multidisciplinary Knowledge Cafe, in which the preliminary work for the development process is done.

RE-DT-UX In their chapter "RE-DT-UX - Moving from Discipline-Based Approach to a Role-Based One," Kerstin Roese, Katharina M. Zeiner, and Rainer Wasgint from Siemens AG discuss the use of design thinking for the development of human-centered user interfaces. In their contribution, they address a central challenge in the development of contemporary information systems, because the design of the user interface will be of great importance in the future. Only if users like to use an application, they will use an information system in the long run. In the traditional "discipline-based approach," usability engineers, design thinkers, and requirements engineers work together as independent experts in the project teams. According to the authors of this chapter, in the future two groups of roles will be distinguished in a software project, roles that deal with the "what" and roles that deal with the "how." This chapter shows how these roles work together to design and implement a good user interface.

Cluster Governance

Change Management Martha Fritsch describes in her contribution "Understanding the introduction of Design Thinking as a change process" with the help of a concrete case study that the combination of requirements engineering and design thinking is not only non-trivial but can also fail. She shows that it takes a well-organized change process to integrate design thinking into a traditional software development department. Among other things, the author takes Lewin's model for organizational change as a basis and shows that there can be resistance to the introduction of design thinking. She states that communication is a central success factor in change management.

Strategic Roadmap Markus Guentert, Holger Rhinow, and Christoph Meinel of the Hasso Plattner Institute describe in their chapter "From Project Plans and Backlogs to Strategic Roadmaps: The Evolution Towards Value-Oriented Thinking in Requirements Engineering" a strategic roadmap that focuses on value, measured primarily in desirability of a solution for the customer. The three authors are of the opinion that classical project planning and also the backlogs in agile approaches are too strongly focused on feasibility. Their proposal for a strategic roadmap is structured according to the creation of customer value.

Managing Tension Dario Staehelin, Mateusz Dolata, and Gerhard Schwabe focus in their chapter "Managing Tensions in Research Consortia with Design Thinking Artifacts" on the use of design thinking to manage tensions in a research project. The goal of the research project was to develop a language-based customer advisor for a bank. Using three concrete examples, operational tensions, decision-making tensions, and solution tensions, they show how the people-oriented approach of design thinking is able to contribute to the resolution of tensions in software projects.

Cluster Use Cases

Platform Design Michael Jakob from the Deutsche Bundesbank and Jennifer Hehn who is one of the editors of this book show in their contribution how design thinking is used in a critical and complex project of the Deutsche Bundesbank to develop an innovative and human-centered application. The project described is about replacing the extranet of the Deutsche Bundesbank with a comprehensive innovation platform for all external partners. In their contribution, Jakob and Hehn show how classical approaches, agile work forms such as Scrum, and design thinking can be combined across all phases of a software project to develop an innovative and human-centered solution. They adapt their approach "prototyping - testing – refinement" to the needs of a large software project. The contribution of Jakob and Hehn can be seen as a blueprint for the future approach in software projects where the needs of many users with different interests have to be considered.

Store Floor Support System Markus Durstewitz from Airbus Operation AG and Thomas Abrell from Volkswagen AG describe in their chapter "Design Thinking in a Large Manufacturing Organization - A Case Study for Designing a Smart Support System for the Store Floor" how design thinking was used in the development of an intelligent support system for employees in production at Airbus. During the identification of the requirements for the information system, the future users were integrated into the development process. The methods of "Needfinding" and "Prototyping and Testing" were used. It is exciting that the third planned step in design thinking "Implementing" was not realized, because the project was integrated into the MES project. The contribution of Durstewitz and Abrell can be classified as a practical experience report.

Digital Platform In his chapter "Digital Platform Design at the Edge of Complexity," Emanuel Stöckli from Esuarance AG deals with the use of design thinking in the development of digital platforms. In doing so, he turns to the topic of reusing software modules to reduce complexity in the development process. Stöckli treads a fine line. On the one hand, it is important to maintain the innovative power of design thinking and, on the other hand, to develop reusable modules in order to increase the efficiency of the development process and reduce costs. The author shows how this fine line is mastered in his companies.

Healthcare/Emergency Department In his chapter "Design Thinking in Healthcare - Enabler for Digitalization in Complex Environments," Christophe Vetterli looks at the use of design thinking in the development of software-intensive ecosystems in healthcare. The use of design thinking in healthcare has gained importance in Switzerland after the Swiss government selected design thinking as a key method for innovation in healthcare. Vetterli uses three case studies to show how design thinking was used as a first step in the development of healthcare software solutions in his professional environment. In all three projects, the human centricity of design thinking plays an essential role.

Smart Personal Assistant In their chapter "It Takes Two to Tango: Design Thinking and Design Patterns for Better Systems Development," Ernestine Dickhaut from the University of Kassel, Andreas Janson and Jan Marco Leimeister from the University of St. Gallen focus on the recognition of design patterns by using design thinking to increase efficiency in software development processes. Specifically, the focus is on developing software for voice assistants such as Alexa and Siri. From the spectrum of design thinking methods, interviews with future users are primarily used. It is exciting to see how design patterns are developed to deal with new data protection guidelines.

If one evaluates the contributions in this book as a whole, it becomes clear on the one hand that the use of design thinking in requirements engineering makes sense. On the other hand, the contributions show that there is still a lot of potential to better combine design thinking and requirements engineering. Many good approaches are recognizable in the contributions, but much integration work will still have to be done in the future. Against this background, the goal of this book has been achieved, namely to present the state of the art in science and practice and to provide an impetus for future research and work in practice.

We thank all authors for taking the time to write a contribution for this book. It was not a matter of course in the time of the pandemic to take time between video conferences from home or conducting virtual lectures to write a contribution for this edited volume. We would like to thank Barbara Brenner and Nadine Barth for their support and Barbara Bethke from Springer Verlag for the unbureaucratic cooperation and for the opportunity to publish this book at Springer Verlag.

We as editors are aware, as mentioned earlier, that this book represents a beginning. Further research in science and activities in practice are necessary to realize the potential of using design thinking in requirements engineering. We welcome all suggestions, contributions to the discussion, criticism, and, of course, positive feedback. Please direct your comments to jennifer.hehn@bfh.ch.

Bern, Switzerland
Karlskrona, Sweden
St. Gallen, Switzerland
Munich, Germany
January 2022

Jennifer Hehn
Daniel Mendez
Walter Brenner
Manfred Broy

References

Hehn J (2020) The use of design thinking for a human-centered requirements engineering approach. Dissertation, University of St. Gallen. Baier Druck, Heidelberg

Hehn J, Mendez D, Uebernickel F, Brenner W, Broy M (2020) On integrating design thinking for human-centered requirements engineering. IEEE Software 37(2):25–31

IEEE (2000) Recommended practice for architectural description of software-intensive systems. IEEE Std 1471-2000:1–30

Contents

About the Authors

Thomas Abrell is a multidisciplinary design thinking practitioner and scholar, currently working as a Strategist at Volkswagen Commercial Vehicles. Prior to engaging with Volkswagen, he was working as Innovation Manager at Airbus, where he was the project lead for the case described here. He holds a Ph.D. in Business Innovation (University of St. Gallen, Institute of Information Management), an M.Sc. in International Design Business Management from Aalto University (Helsinki), and an M.A. in Design (Tongji University, Shanghai).

Walter Brenner joined St. Gallen University in 2001 as a professor after having held the Chair of Information Systems at the University of Essen (Germany) for two years, and at Technical University of Freiberg (Germany) for 7 years. Currently, Professor Brenner acts as Director of the Institute of Information Management. He published more than 300 articles and more than 35 books. His research focuses on industrialization of information management, management of IT service providers, innovation and technology management, and management of artificial intelligence.

Professor Brenner received a graduate degree in business administration (lic. oec.) and a Doctorate (Dr. oec.) from the University of St. Gallen. Prior to joining academia, Professor Brenner worked as Head of Application Development with Alusuisse-Lonza AG (Switzerland).

Manfred Broy's research is in software and systems engineering in both theoretical and practical aspects. This includes system models, specification and refinement of system and software components, specification techniques, development methods, and verification. One of the main themes is the role of software in a networked world. Under Manfred Broy's leadership, as a member of acatech, the study Agenda Cyber-Physical Systems was created for the Federal Ministry of Research to comprehensively investigate the next stage of global networking through the combination of cyberspace and embedded systems in all their implications and potentials.

Ernestine Dickhaut is a Ph.D. student and research associate at the Department of Information Systems and the Research Center for Information System Design (ITeG) at the University of Kassel. Her research focuses on the codification of conflicting, domain-specific knowledge, and how this can be made accessible to system developers.

Mateusz Dolata is a postdoctoral researcher at the Department of Informatics, University of Zurich. His main areas of expertise include design thinking, human–AI collaboration, and IT use in frontline services. He applies a multidisciplinary perspective shaped by his background in computational linguistics, philosophy, and applied computer science. His research has appeared in journals and proceeding series in Computer-Supported Cooperative Work and in Information Systems.

Markus Durstewitz has 30 years of work experience in the aviation business. As Head of Design Thinking at Airbus, he is responsible for developing an effective innovation framework and ecosystem that delivers value to customers and users. His special focus is on digital transformation and data-driven services offering new ways of collaboration along the complete value chain of aviation in product-service design as well as in operations. He holds a Ph.D. in Cognitive Engineering (University GhK Kassel, Man-Machine Systems Laboratory in collaboration with Euricso at ISAE, Toulouse) and a Diploma in Aerospace Engineering (Technical University Stuttgart).

Michael Felderer is a professor at the Department of Computer Science at the University of Innsbruck, Austria, and a guest professor at the Department of Software Engineering at the Blekinge Institute of Technology, Sweden. His fields of expertise and interest include software quality and testing, software processes, data-driven software engineering, requirements engineering, and empirical methods in software engineering. Michael Felderer performs his research in close collaboration with companies and has more than 15 years of industrial experience as a senior executive consultant, project manager, and software engineer. For more information, visit his website at mfelderer.at.

Martha Fritsch (née Jagoda) studied Media and Information Studies with a focus on Management at Offenburg University of Applied Sciences. She then worked as an academic assistant in the field of usability engineering and as a lecturer in corporate communications, marketing trends, international marketing, and innovation management at Offenburg University of Applied Sciences. Martha Fritsch completed her academic career at the end of 2014, initially at the Humboldt University in Berlin, with a doctorate in internal brand management. Since then, she has worked in the fields of usability engineering, agile software development, corporate communication, and innovation management. Design thinking has always accompanied her, but the introduction of design thinking as a method has been a challenge on several occasions. For this reason, she dealt in depth with the topic of change management within her professional career. She is currently Head of User Experience & Conception at an agency for digitalization.

Daniel Gadner completed his master's studies at the University of Innsbruck in Information Systems. He is interested in the integration of design thinking elements in agile software development frameworks like Scrum. Currently, he takes the role of a product owner at the ecommerce company Internetstores where he is responsible for further development of web shops like fahrrad.de or addnature.com. Besides requirement analysis, refinement, and prioritization, he also coordinates and organizes the respective implementation activities in an agile web development team.

Markus Guentert is an alumni of the Hasso Plattner Institute and a design thinker at heart. He started his career as a digital product strategist in a large German consultancy where he shaped key innovation initiatives of corporates and SMEs from discovery to MVP launches. Now working as an independent consultant, he mainly focuses on organizational and product-related coaching. So far, he has worked with more than 15 major clients in the DACH region, primarily in financial services, digital health, and mechanical engineering.

Jennifer Hehn is professor at the Institute for Digital Technology Management, Bern University of Applied Sciences, and associate lecturer at the University of St. Gallen, Switzerland. Her research interest is focused on combining human-centered design and agile innovation principles to develop innovative software-intensive solutions. Jennifer held various positions in the context of design thinking and innovation, for example as innovation manager at Deutsche Bundesbank or as a senior manager at the innovation consultancy IT Management Partner St. Gallen AG (ITMP). She has led numerous innovation projects within companies of all sizes across various industries, including software, pharmaceuticals, banking, and insurance. Until 2017, Jennifer was the Executive Director of the Design Thinking program at the University of St. Gallen and global coordinator of the SUGAR Network, a global network for design thinking with partner universities worldwide.

Michael Jakob is senior IT project manager at Deutsche Bundesbank, the central bank of Germany. He leads strategic IT projects in the area of digitalization and supports teams to develop and implement new services and solutions, e.g., video conferencing, Intranet, and eBusiness platform. He is a certified Design Thinking Coach (HPI), Professional Scrum Master, Professional Product Owner, Project Management Professional (PMP), and ITIL Expert. Additionally, he has been supporting the introduction of design thinking and Scrum in the working environment of Deutsche Bundesbank and the transformation of IT since 2018. From 2007 until 2012, Michael was responsible manager for Mainframe Infrastructure operations and engineering. During this time, he also supported international projects of the European System of Central Banks when introducing high-available payments systems (Target2, Target2-Securities). Currently, his professional focus is on applying and combining different human-centered and agile frameworks to develop, implement, and launch user-centric and secure IT solutions at Deutsche Bundesbank. At the moment, he coordinates the renewal of the extranet platform.

Andreas Janson is a Postdoctoral Researcher at the Institute of Information Management (IWI-HSG) at the University of St. Gallen, Switzerland. He obtained his Ph. D. from the University of Kassel, Germany. His research focuses on service design, smart personal assistants, decision-making in digital environments, and digital learning. His research has been published in leading information systems and management journals such as the *Journal of the Association for Information Systems*, *European Journal of Information Systems*, *Journal of Information Technology*, and *Academy of Management Learning & Education*.

Kim Lauenroth designs digital solutions since 2011 and heads the Competence Center for Requirements Engineering at adesso SE since 2013. He studied computer science, business administration, and psychology at the Technical University of Dortmund and received his Ph.D. in product line engineering from the University of Duisburg-Essen.

At Bitkom—Germany's digital association, he is committed to establishing digital design as a design profession for digitalization. Furthermore, he is active in the International Requirements Engineering Board for the standardization of professional education and training in requirements engineering and digital design.

Jan Marco Leimeister is a Full Professor and Director at the Institute of Information Management at the University of St. Gallen. He is also Full Professor and Director of the Research Center for Information System Design (ITeG) at the University of Kassel. His research covers digital business, digital transformation, service engineering, and service management. His research has been published in leading journals such as *Information Systems Research*, the *Journal of Management Information Systems*, *Journal of the Association for Information Systems*, and *Journal of Information Technology*.

Michael Lewrick has worked very intensively with the mindset, which enables us to solve different types of problems. He is the author of the international bestsellers *Design Thinking for Business Growth*, *The Design Thinking Toolbox*, and *The Design Thinking Playbook*, in which he describes the mindful transformation of people, teams, and organizations. He works intensively with universities and companies and places the self-efficacy of people, in personal and organizational change projects, at the center of his activities. During the last 10 years, he has extended his toolbox to the design of data and business ecosystems. As an internationally recognized expert in the field of digital transformation and the management of innovations, he was engaged in the realization of many projects, initiatives, and research projects globally.

Christoph Meinel (*1954) is CEO and Scientific Director of the Hasso Plattner Institute for Digital Engineering gGmbH (HPI) at the University of Potsdam.

Christoph Meinel is CEO and Scientific Director of the Hasso Plattner Institute for Digital Engineering gGmbH (HPI). He is also Vice Dean of the Faculty of Digital Engineering at the University of Potsdam. Christoph Meinel holds the chair of Internet Technologies and Systems. He is engaged in the fields of cybersecurity and digital education. He has developed the MOOC platform openHPI.de, supervises numerous Ph.D. students, and is a teacher at the HPI School of Design Thinking, where he is also scientifically active in research. Earlier scientific work concentrated on efficient algorithms and complexity theory.

Christoph Meinel is author or coauthor of more than 25 books, anthologies, ands numerous conference proceedings. He has had more than 550 (peer-reviewed) papers published in scientific journals and at international conferences and holds a number of international patents. He is a member of the National Academy of Science and Engineering (acatech), director of the HPI-Stanford Design Thinking Research Program, honorary professor at the TU Beijing, visiting professor at Shanghai University, concurrent professor at the University of Nanjing, and member of numerous scientific committees and supervisory boards.

Daniel Mendez is full professor at the Blekinge Institute of Technology, Sweden, and Lead Researcher heading the research division Requirements Engineering at fortiss, the Research and Transfer Institute of the Free State of Bavaria for software-intensive systems and services. After studying Computer Science and Cognitive Neuroscience at the Ludwig Maximilian University of Munich, he pursued his doctoral and his habilitation degrees at the Technical University of Munich. His research is since then on empirical software engineering with a particular focus on interdisciplinary, qualitative research in requirements engineering, and its quality improvement—all in close collaboration with the relevant industries. He is further editorial board member for EMSE and JSS where he co-chairs the special tracks Reproducibility & Open Science (EMSE) and In Practice (JSS), respectively. Finally, he is a member of the ACM, the German Association of University Professors and Lecturers, and of the International Software Engineering Research Network. Further information is available at http://www.mendezfe.org.

Holger Rhinow is the Chief Product Officer for a new digital learning platform that is being developed at the HPI Academy. Over the last 10 years, Holger consulted more than 100 companies in implementing design thinking and agile frameworks into existing organizational landscapes. Between 2011 and 2018, Holger held a scholarship at the HPI-Stanford design thinking research program.

Kerstin Roese is Head of the User Experience Group Germany at Siemens Technology. She joined Siemens Corporate Technology in 2011 as Expert for Industrial UX and worked a long time as Senior Key Expert for UX methods and processes.

Her professional experience is built on study of psychology at Humboldt University Berlin with a Master's in Engineering Psychology, Pedagogical Psychology, and Clinical Psychology and a well-recognized and award-winning Ph.D. work in System and Mechanical Engineering at the Technical University Kaiserslautern. After her Ph.D., she worked several years as Professor for user-centered product development at the TU Kaiserslautern. After a visiting professorships at Copenhagen Business School (DK) and at the Industrial Design School at KAIST (KOR), she stopped her scientific carrier to come back to agile project work and to start her work as UX influencer at Siemens. In 2018, she founded the UX CAMP at Siemens—an ongoing initiative to push the human-centered mindset at Siemens. Her publication record has over 80 articles and paper and she is a well-received keynote speaker at conferences and discussion partner in several industrial forums. Kerstin is the founding president of the German UX professional association (GC upa) and active in other UX-related communities and workstreams. With her background, she is always interested in innovation for work methods and new impulse for digital user experience.

Gerhard Schwabe has been a full professor at the University of Zurich since 2002. He has studied collaboration at the granularity of dyads, small teams, large teams, organizations, communities, and social networks. In doing so, he follows either an engineering approach ("design science") or an exploratory approach—frequently in collaboration with companies and public organizations. Currently, his research interests focus on blockchain applications and human–robot collaboration.

Dario Staehelin is a Ph.D. student and research assistant at the Department of Informatics, University of Zurich. He holds an M.A. in Business Innovation from the University of St. Gallen. His research interest lies in human–AI collaboration in dyads, small teams, and organizations.

Emanuel Stoeckli is Platform Product Manager at the Zurich-based InsurTech company esurance, where he plays a key role in the design and development of a B2B2C platform. After studies in computational sciences and economics in Zurich and Uppsala, he graduated in 2015 from the University of Zurich as M.Sc. in Informatics, followed by a Ph.D. in Management from the University of St. Gallen (HSG), where he conducted research in collaboration with Allianz Technology in Munich. During his doctorate, he held positions as Research Associate and Design Thinking Coach at the Institute for Information Systems at HSG, and he was a Visiting Researcher at the Center for Design Research (CDR) at Stanford University. Dr. Stoeckli's research focuses on digitalization in the insurance industry, design methodologies, and digital work with a particular focus on feedback exchange. His research has been published in academic journals such as *Electronic Markets* and *Journal of Business Analytics* and in numerous peer-reviewed conference proceedings.

Christophe Vetterli's formal academic training was obtained at the University of St. Gallen (HSG). He holds a Master's in Business Innovation and a Ph.D. in Embedding Design Thinking (which was obtained in strong collaboration with Stanford University). Vetterli is CoFounder and Managing Partner of a leading healthcare consultancy focusing on transformation of healthcare by using design thinking in hospitals and clinics as innovation approach for processes, strategy, and construction projects. He additionally lectures at several international universities and publishes regularly at the intersection of innovation and healthcare.

Rainer Wasgint currently works as a Technical Project Lead and Requirements Engineer at Siemens AG, Technology in Research in Digitalization and Automation in the Technology Field Software & Systems Innovation within the competence group User Centered Experience Design located in Munich. Rainer does research in projects using Scaled Agile SW Developments frameworks like SAFe and is a leading SAFe-certified expert and within Siemens senior key expert. His expertise and special interests lie particularly in the areas of software system families and modeling with long background, UML/SYSML, and agile (system) requirements engineering. His work experience comprises numerous project leadership positions, program management of Siemens programs, and hands-on project work as technical product owner in agile projects. Rainer also contributed to several EU research projects like ESAPS, CAFÉ, and FAMILIES. One of his publications also contributing to a patent was System Decision Making and Operational State Optimization—SDMOSO' which is based on former research in the area of project engineering and contributed to several internal publications leading to the material published here.

Katharina M. Zeiner is Design Thinker and User Researcher at Siemens T RDA SSI UXD-DE. In her work she focuses on the structures and processes we need to put in place to allow teams to do their best work. This ranges from frameworks for Agile UX to UX Maturity assessments. Katharina has a background in psychology (St Andrews, Scotland) and started her professional career studying 3D and lightness perception (St Andrews, Scotland, and TU Berlin, Germany). During this time she realized that she wanted to focus more on the experiences we can create for people. She did so by investigating the factors contributing to positive UX in work contexts (Stuttgart Media University, Germany). She has published extensively on this and has created easy-to-use methods that allow even inexperienced teams to take first steps toward creating products with better UX. Since joining Siemens her focus has shifted slightly from methods that can be used by teams to asking how we might best enable teams to create those experiences by addressing how they work.

Interview with Manfred Broy and Walter Brenner About Design Thinking and Requirements Engineering

Walter Brenner and Manfred Broy

State of the Art

Mr. Broy, what do you understand as "requirements engineering" and where do you see a need for further development?

The task of requirements engineering is simple to outline: It is about defining all requirements for the result of a development project. We focus on software-intensive systems, especially with regard to their functionality, but also quality characteristics and additionally further specifications and, if necessary, also restrictions for the development process and certain realization details. It is obvious that this is of central importance—both for the usability of a system to be developed and for the costs of the development. Here, the critical questions are always the same: Which properties of the system to be developed are the ones really needed? How can often vague and abstract objectives be translated into concrete requirements, for example, in terms of functionality, while exploiting all the potential for innovation? How do you prove that all requirements have been met for the final product? Typically, we distinguish between the question of whether requirements were correctly elicited and whether the captured requirements were correctly implemented. Checking these two properties leads to validation, on the one hand, and verification, on the other hand. Equally important is the question: Do we end up with the optimal requirements for the task, i.e., the optimal system properties? Typically, when building such systems, the idea of the developers is often strongly determined by reproducing existing solutions with the help of software. Since software provides a huge solution space

W. Brenner (✉)
Institute of Information Management, University of St. Gallen, St. Gallen, Switzerland
e-mail: walter.brenner@unisg.ch

M. Broy
Department of Informatics, Technical University of Munich, Munich, Germany
e-mail: broy@in.tum.de

J. Hehn et al. (eds.), *Design Thinking for Software Engineering*, Progress in IS,
https://doi.org/10.1007/978-3-030-90594-1_1

and, in particular, opens up completely different possibilities, it is necessary to break out from traditional ideas and enter innovative solution spaces. This gives rise to central challenges in requirements engineering for which there is a strong need for further development.

Mr. Brenner, what do you understand as "Design Thinking" and where do you see a need for further development?

Design Thinking emerged in the 1960s and 1970s at major American technical universities, such as MIT, Stanford University, and Carnegie Mellon University. In my view, Design Thinking is primarily a way of thinking, almost a culture, that puts people at the center when overcoming challenges or developing new ideas. It is mainly about discovering obvious and hidden needs of people and taking them as a starting point for the later solution. These needs can then become requirements for a software product. Another important component of the Design Thinking approach is the construction of prototypes that are as concrete as possible and that can be tested by later users. These tests should be done as early as possible according to the often-cited principle of "fail fast, fail early." These tests can serve to validate a solution, i.e. they provide answers to the questions of whether the correct system is being built. In the past decades, many methods have emerged in the Design Thinking environment to help projects take a human-centered approach, build prototypes, test prototypes, or identify obvious or hidden human needs. There is definitely a need for further development of Design Thinking. In recent years, a clear trend towards digitalized (i.e., software-intensive) solutions has emerged in the implementation of projects with Design Thinking. Design Thinking is, however, technology-agnostic. Nevertheless, as a result of many projects in which Design Thinking is used, software comes out as a result. Technology neutrality at the beginning of a project always leads to unconventional, disruptive solutions, even when it comes to ideas for new software solutions. However, technology neutrality is also the cause for large "frictional losses" that can be seen in the implementation of innovative solutions created with Design Thinking, if said innovative ideas are implemented in software. In my opinion, it is, therefore, very important to align Design Thinking more strongly with the challenges of digitization. Specifically, I refer to focusing Design Thinking more strongly on the development of solutions that result in software.

Opportunities

Mr. Broy, where do you see opportunities for IT projects when the spectrum of methods and tools used in requirements engineering is expanded to include Design Thinking?

As already mentioned above, one of the great challenges in system development tasks, which are often characterized by outdated ideas of solutions, is to broaden one's view and to recognize what other possibilities are there and what the users of a projected system really need. Let's take the example of navigation: If we consider the navigation situation a few decades ago, navigation was essentially done with the help of paper-based land and roadmaps. A naive implementation in software-intensive systems would lead to displaying such land and roadmaps in the computer and perhaps creating possibilities for zooming into such maps very precisely in order to display all details for which traditionally different maps were required. As we all know, today's navigation systems work in a completely different way: they take care of route determination and directly support the selection of alternatives in the choice of a route. Even during the current journey, they provide information at the right time, for example, on when to take a turn within the route. This could be supported even better with augmented reality. Autonomous driving is then a consequent next step, in which the vehicle drives the route independently. This example shows how difficult, but also how important, it is in a system development to understand exactly which options are available for the functionality of the system based on the diverse possibilities of software.

Mr. Brenner, where do you see opportunities for Design Thinking projects if Design Thinking and requirements engineering are better combined?

Design Thinking has traditionally been focused on physical products. Over the last 15 years, the focus of innovation has shifted. Increasingly, digital innovation has come to the fore, as demonstrated by the example of making navigation easier. Design Thinking can help identify the system characteristics that subsequent users expect from navigation software, while also contributing to making the software human-friendly to use. That way, Design Thinking can make a contribution to meeting the challenge of "digitization." However, we have seen that in many projects, the implementation of ideas in software, which is used in Design Thinking projects, is not free of problems. Many innovative ideas that emerge through Design Thinking are not documented in a way that enables software developers, who implement the results of projects that use Design Thinking, to understand and implement important components. It is very unsatisfactory when a good idea is changed during the implementation in software in a way that the actual innovation is lost. Based on these experiences, it is an important further development for me to complement Design Thinking with methods, tools, and experiences of requirements engineering. The goal must be to work out innovative ideas so that the "friction loss" during the implementation in software is as low as possible.

Deficits

Mr. Broy, where do you see deficits in IT projects today, especially when it comes to requirements engineering?

The deficits are manifold. On the one hand, requirements are often not formulated cleanly, or worse, there is no careful exploration of which requirements are appropriate for the problem. This results in systems that, on the other hand, do not contain important requirements, do not open up innovative possibilities, and, in the worst case, are incorrect with respect to requirements that have already been found or implement requirements that are not relevant at all. In practice, this often means that extensive learning processes take place during a system development. Sometimes new insights into which functionality is actually needed only emerge after the system has been delivered, which then has to be laboriously inserted into the systems through changes. In agile development, one tries to counter this by defining requirements in large parts only in the course of the project, when the learning process is already sufficiently advanced. A more basic attitude is to be ready to make fast and constant changes to the requirements. But whether this guarantees an efficient approach is another question. Design Thinking and related techniques can help here by trying to detach the learning process as much as possible from the question of implementing specific implementation details and to anticipate them in independent, upstream projects so that innovative solutions are not neglected with much greater certainty.

Mr. Brenner, where do you see deficits in Design Thinking projects when it comes to digital innovations?

Before I talk about the deficits of Design Thinking, let me briefly discuss the deficits of requirements engineering that you mentioned and how Design Thinking can help to compensate for them. An advantage of Design Thinking lies in technology neutrality, at least at the beginning of an innovation project. It makes it possible, together with later users, to go through their requested learning process openly and as free of restrictions as possible and to begin evaluating implementation alternatives only when it is clear what the solution should look like. This saves all parties involved from having to recognize whether the solution or software "fits" or not only when the fully developed solution or software is available. I see the biggest shortcomings of Design Thinking in the difficult interface between Design Thinking and software engineering. Design Thinking is a very "soft" method that focuses on people and primarily relies on qualitative methods, deliberately neglecting aspects of implementation. It is about innovative, people-oriented ideas and not primarily about implementable solutions. This way of thinking is the great advantage of Design Thinking. It enables disruptive innovations. However, we see that this "remoteness from implementation" in the development of the innovative ideas, overwhelms the people who then have to implement them, especially software specialists. The innovative solutions of Design Thinkers are not formulated in the language of computer scientists, in particular software engineers, and sometimes the "crux" of the innovative solutions is hidden in impressive prototypes that are, however, practically undocumented. In this context, it is increasingly necessary to incorporate the computer scientists' computer science-oriented way of thinking and their methods and tools for defining requirements into the world of Design Thinking. However, it must be ensured that the deliberately "non-formal" character of Design Thinking does not get lost.

Mr. Broy, what does human-centeredness mean in requirements engineering today and how important will it become for IT projects in the future?

Human-centric design is an absolute "must-have" in light of the fact that today's software-intensive systems accompany us extensively in our everyday lives. This requires a clear understanding of the situation in which people use certain systems and functionalities, and how the functionality of the systems is delivered to people so that they can optimally support them in their tasks. The use of a software system always follows a specific purpose and takes place in a specific context of action. This also requires that there is coordination between the system functions and the cognitive processes as well as the physical processes that take place at the user. The coordination is then reflected in the functionality, but also in the design of the human–machine interaction and leads to the central concept of user experience (UX), which aims to ensure that a system optimally meets the requirements and expectations of its users and delights them in its use.

Mr. Brenner, what does human-centeredness mean in Design Thinking and what methods does Design Thinking provide?

Design Thinking helps to identify the purpose and the context of action of a solution. A central guiding principle of Design Thinking is: "Innovation is made by people for people." The part "for people" is action-guiding for the approach in Design Thinking. One of the central aspects is the identification of obvious and latent needs as the starting point of innovative ideas or the solution of problems. Everyone knows this from their own life. Design Thinking offers numerous methods to get very close to people's needs and the context of action. For example, there are special methods to conduct interviews. For me, shadowing is an excellent example. Potential future stakeholders of the solution are accompanied by a specially trained person who observes and documents exactly what the stakeholder is concerned about. The new insights that are gained as a result are tremendous. Another impressive aspect of human-centeredness in Design Thinking, in terms of validating the later solution, is the early radical testing of new solutions with affected people, described by the phrase "fail fast, fail early." From the very beginning, Design Thinking aims not to develop written concepts, but rather prototypes that are as tangible as possible, which are intended to show what the planned innovative idea or problem solution could look like. Partly the prototypes consist of paper and sometimes videos or role plays show the solution. These prototypes are tested with affected people as early as possible in the innovation process. Their feedback radically influences the further development of solutions without compromise. If a prototype is rejected during testing, the solution "dies." If a prototype meets with approval, it is developed further. That is the core of Design Thinking. It is actually a cleverly designed cyclical process of identifying needs, building, and testing prototypes, and mercilessly considering the results of those tests.

State of the Practice

Mr. Broy, what is the state of practice today when it comes to using Design Thinking in IT projects?

I think there are big differences here. In many projects, Design Thinking methods are hardly used at all. However, some companies are already much further along and routinely use Design Thinking and other exploration and creativity techniques. All in all, a rethinking process must begin here, a process of realization of how important and indispensable such approaches are.

Mr. Brenner, what is the state of practice today when it comes to translating ideas from Design Thinking projects into software?

Design Thinking has become widespread as a way of thinking, problem-solving, and innovation method in many companies in recent years. Within just a few years, Design Thinking has gone from being an exotic fringe phenomenon to a central component of innovation management. Design Thinking also performs outstanding services in the context of digitization and digital transformation. Numerous user-friendly websites or apps and many successful digitized products have been created with the help of Design Thinking. Nevertheless, there are opportunities to better use the potential of Design Thinking for the development of digital products and services. In practice, the innovation process that uses Design Thinking and the implementation process, specifically software engineering, is still too often too far apart. A spasmodic search is made for an interface between the results of Design Thinking, usually prototypes, and the subsequent software development process. This is exactly where we need to start in the future. In my opinion, companies that integrate both worlds could produce great innovative digital products and services.

Skills and Competencies

Mr. Broy, what additional skills will people need in the future who are involved in developing human-centered software?

As already indicated above, tailoring systems to people's requirements, i.e., developing human-centered software, means gaining a precise understanding of the relevant people's action situation and of how they are going to be supported by the systems. This requires a great deal of knowledge about people, their cognitive abilities, their sensitivities, their wishes, values, expectations, and goals. Especially at the beginning of the information age, software-based systems were often built where people had to adapt to the requirements of the systems and not vice-versa. Hence, in the requirements elicitation and system development processes, we have to focus on the ability to grasp exactly in which situation people use a system and what is required for this.

Mr. Brenner, what additional skills will traditional design thinkers need in the future if more and more innovation results in software?

The ideal competency profile for a traditional design thinker corresponds to the T-profile. The horizontal part of the T corresponds to the breadth of competencies for a design thinker, the vertical part of the T represents an area of knowledge in which the design thinker has in-depth knowledge. However, this profile does not only correspond to the ideal conception of the design thinker. If truth be told, this is how most people's competency profile is structured, although of course there are large differences in the breadth and depth of knowledge. In recent years, numerous new areas of competence have been added, both in terms of breadth and depth. Today, if you want to solve central problems of economy and society, you need basic knowledge in ecology, ethics, gender issues, cybersecurity, data protection, artificial intelligence and of course software engineering. Therefore, every design thinker should have basic knowledge in software engineering and requirements engineering. They are also required to acquire basic knowledge in the important areas of digital transformation. In terms of depth, today almost every project requires people with a high level of competence in information technology and software development. People who have this knowledge must be integrated into the project teams.

Challenges

Mr. Broy, what challenges do you see for traditional computer scientists if they engage in Design Thinking?

I am not entirely clear on what a "traditional" computer scientist is. Does it refer to the way computer scientists were trained at the beginning, in the first courses of study in computer science? At that time, the focus was—for understandable reasons—on getting to know the most important concepts of computer science in the narrow sense. These were aimed at understanding the structure of computer science systems: The forms of implementation, the different aspects in algorithms or programming.

However, this is completely detached from the questions already discussed above. How do I achieve building systems that respond optimally to the expectations, demands, and sensitivities of the users? That is a completely different question. It is detached from computer science details at the first moment; on the other hand, it is closely related to them, since certain ideas about how functionalities of systems should look like can of course not be considered detached from the question of how to implement such functionalities efficiently and effectively. Computer scientists must have this understanding today, and then approaches such as Design Thinking will show valuable ways to implement this practically.

The result of a Design Thinking process is, on the one hand, a prototype and, on the other hand, a much deeper understanding by all participants of the requirements for the system to be developed. The understanding is perhaps the most valuable result, but it also immediately introduces a challenge: How can this understanding subsequently be transported to the developers of the system, in particular also to the software engineers? This applies equally to the prototype: With a prototype, it is of course not clear which features of the prototype are rather coincidental and thus irrelevant for the successful implementation in the and which approaches in the prototype are crucial for the applications. In a nutshell: It is a challenge to identify, implement, and record the respective essential requirements from a prototype so that they can find their way into the software development process. For this very reason, we believe that it is important to build a bridge between Design Thinking and requirements engineering.

Mr. Brenner, what challenges do you see for classic design thinkers if they become more involved with requirements engineering and software engineering?

After working with Design Thinking and traditional design thinkers for many years, I see primarily "human" challenges. Design thinkers have to respect that digital solutions, i.e., software, are the result of many innovation projects and that they can only leverage the potentials of Design Thinking if they fully use the potentials of computer science and if there is as little "friction" at the interface between Design Thinking, requirements engineering, and software engineering as possible. This means that design thinkers respect requirements engineering and software engineering methods and tools, and do not refer to people who are at home in that world as "technocrats" and treat them with disdain. Design thinkers need to accept that it takes another step to implement in software. Initial conversations with "real" design thinkers have shown me that they are, to put it mildly, not very enthusiastic about being asked to use more formal documentation methods. They counter that they would feel constrained and that the character of Design Thinking projects is changed in the wrong direction. However, I don't see an alternative to extending Design Thinking to include aspects of computer science and especially requirements engineering. Digital innovations are still gaining importance for the development of innovative solutions for central challenges for economy and society. Only if we succeed in incorporating more computer science into Design Thinking, the potentials of Design Thinking can be used optimally.

Goals for the Book

Mr. Broy, why are you participating in this book and what can a reader learn from it?

It has a lot to do with my personal experience in dealing with requirements. In the first few years when I was dealing with this, the focus—to me—was on how to formulate requirements so that they are clearly described and can subsequently be used in system development and also in system verification. The question of which functionalities users really need in which form was rather in the background. Today, my view has almost reversed: It is crucial to first understand which functionalities users need and then to address the question of how to model and analyze it to be able to implement it optimally. The field of requirements engineering is still very fragmented. There are only a few methods that are generally accepted or even standardized. This is also reflected in the tools that are important for practical implementation. The tools are much weaker and less suitable for development tasks than would be possible in principle today. Only the merging, combining, and standardizing of approaches can lead to the emergence of a more standardized field by understanding how the different methods and approaches intertwine. This is precisely the goal of this book.

Mr. Brenner, why are you participating in this book and what can a reader learn from it?

For me, this book is a first and important step in connecting Design Thinking, requirements engineering, and software engineering. This book builds bridges between the worlds of computer science and Design Thinking from the perspective of both science and practice. Readers of this book will get an overview of the state of the art in science and practice as well as valuable advice on how to use Design Thinking and requirements engineering in combination in projects. I am very grateful that we are able to produce this book in cooperation with members of the Faculty of Computer Science and fortiss at the Technical University of Munich, members of the Institute of Information Management at thc University of St. Gallen and the Blekinge Institute of Technology in Karlskrona in Sweden and numerous other personalities from science and practice. I would like to thank all those involved for their cooperation and the time they have invested in the contributions.

Combining Design Thinking and Software Requirements Engineering to Create Human-Centered Software-Intensive Systems

Jennifer Hehn and Daniel Mendez

Introduction

The success of any software-intensive system anchors in the question of how well it reflects its users' needs (Maguire and Bevan 2002) and surrounding constraints. "Getting the requirements right"—which is often associated with the term *Requirements Engineering*—is consequently seen as one of the most significant endeavors in development projects (Broy 2006; Robertson and Robertson 2013). It is typically associated with the initial phases of a software development life cycle and its major aim is to decide upon the relevant functional and nonfunctional properties of software-intensive products. Elementary tasks toward this goal include the elicitation of requirements, their analysis and negotiation also in terms of reaching consensus among all relevant stakeholders, their specification to accommodate subsequent engineering activities, and their validation in terms of ensuring the requirements' quality (e.g., correctness and consistency, among other attributes).

In accordance with the terminology introduced by the International Requirements Engineering Board,[1] Requirements Engineering, therefore, denotes the "systematic and disciplined approach to the specification and management of requirements with the goal of understanding stakeholders' desires and needs and minimising the risk of

[1] See the IREB glossary, available at www.ireb.org

J. Hehn (✉)
Institute for Digital Technology Management, Bern University of Applied Sciences, Bern, Switzerland
e-mail: jennifer.hehn@bfh.ch

D. Mendez
Blekinge Institute of Technology, Karlskrona, Sweden

fortiss GmbH, Munich, Germany
e-mail: daniel.mendez@bth.se

J. Hehn et al. (eds.), *Design Thinking for Software Engineering*, Progress in IS,
https://doi.org/10.1007/978-3-030-90594-1_2

delivering a system that does not meet these desires and needs." Given the human-centric nature of software—in the end, software is made by humans for humans—Requirements Engineering is undoubtedly a critical determinant for software quality, regardless of how Requirements Engineering exactly manifests itself in practical settings.[2] In a world pervaded by software and where most of our daily routines are supported—if not dominated—by software-intensive systems, excellence in RE is a de facto key. At the same time, many companies struggle with capturing the users' needs effectively, often leading to software-intensive systems which (1) either miss important requirements, (2) reflect incorrect requirements (or incorrect assumptions), or (3) which reflect—technically speaking—the correct functionality, but are still rendered unusable as they lack important nonfunctional properties from the perspective of their end users. This gives rise to the need for new approaches that allow for a more human-centered Requirements Engineering.

In the following, we first elaborate on the difficulties and limitations of contemporary Requirements Engineering principles and approaches, before motivating their symbiotic relationship with Design Thinking to create software-intensive systems in such human-centered way. Exploring the relationship of both historically grown worlds as part of an integrated approach is in scope of this book chapter.

Requirements Engineering and its Limitations

Many companies struggle in the complex endeavor of establishing a high-quality Requirements Engineering and, in consequence, many projects suffer from insufficient Requirements Engineering. One of the key characteristics of Requirements Engineering is its volatile nature and its sensitivity to its practical context. Many things are not clear at the beginning of a project and a methodology, method, or tool that might fit very well the needs of one project could be completely alien to the needs and the culture of the next. This is what renders Requirements Engineering as something hardly standardizable with universal one-size-fits-all solutions and, thus, a discipline difficult to master. It is, therefore, not surprising that 33% of software development errors are estimated to have their origin in insufficient Requirements Engineering (Emam and Koru 2008; Mendez Fernandez and Wagner 2014). Moreover, 36% of those errors are known to lead to project failures. Requirements Engineering is therefore not only difficult to handle, but it is also crucial for project success. Further studies corroborate the criticality of Requirements Engineering as they show how requirements errors may represent 40% of the total project costs; it is

[2] We can often observe that RE is subsumed under the umbrella of software process models or product management approaches, often without using the term "Requirements Engineering." In this chapter, we do not distinguish between those various approaches but refer to the handling of requirements—from their inception to their specification and validation—which is in scope of any product development regardless of the chosen approach and terminology and regardless of whether it is done explicitly or implicitly.

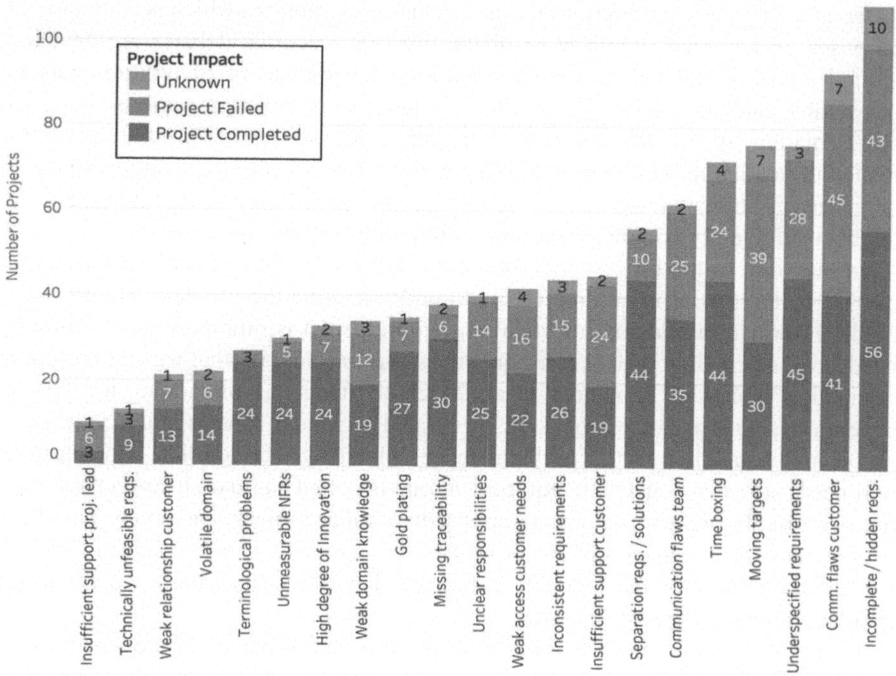

Fig. 1 Top 21 requirements engineering problems as revealed by the "Naming the Pain in Requirements Engineering" initiative. See also www.napire.org for more information including publications and open data sets

commonly accepted that when these errors are found late in the development process, their correction can make up to 200 times more than when correcting them during in early development stages (Venkatesh Sharma and Kumar 2013).

We initiated, at the time of writing this chapter already a decade ago, a globally distributed, biyearly replicated family of surveys to gain insights into contemporary practices and challenges in Requirements Engineering: The Naming the Pain in Requirements Engineering initiative (short: NaPIRE, see also Wagner et al. 2019, Mendez Fernandez et al. 2016).[3] Among the insights we gained are a clearer understanding about the most frequently occurring and most critical problems companies experience, as well as their root causes and their effects (going beyond a binary view on project failure and success). Here, we discovered that a large share of problems is related to human factors (see Fig. 1) and the lack of expertise to deeply penetrate the problem space—and this is regardless of the software process model employed such as "agile" (Mendez Fernandez et al. 2015).

This is not surprising given that Requirements Engineering is historically grown out of engineering disciplines and corresponding worldviews and it involves many

[3] See also the project website NaPiRE.org for further information and related empirical data sets.

different approaches, methods, tools, and techniques—none of which is suited for all purposes. In any case, while most of the primary academic debates are centered around questions related to the specification and refinement of requirements to measurable and, in particular, verifiable requirements covering various forms of representation (for models and natural language descriptions) as well as questions related to facilitating seamless modelling and the transition to the solution space in engineering, little attention is paid to eliciting the actually relevant requirements to obtain a sufficiently complete and correct requirements specification.

In fact, one of the biggest lies we tend to tell ourselves in Requirements Engineering is that the information relevant to understanding the problem space (stakeholder information, context information, requirements) is omnipresent and simply needs to be elicited. A typical consequence of this problem is what we call "solution orientation" (Mendez Fernandez et al. 2012), the tendency of moving too fast to develop a solution and of focusing on related technical aspects often without a proper understanding of the problem to be solved by that solution. Here, actual user needs are often neglected, requirements are invented based on incorrect assumptions or blindly reused from other supposedly similar projects and solely based on the requirements engineer's intuition, or they lack creativity (see Inayat et al. 2015 as well as the results of the NaPiRE initiative). This underlines the need for more problem-oriented ways of thinking.

In fact, today's complexity growth in product development where system and domain boundaries become more and more fuzzy and where human factors become more and more important makes explicit the need for a shift even in RE itself (and corresponding roles and responsibilities) from often technology-centric ways of thinking, tasks, and domain-expertise to problem-centric ways of thinking, mediation, empathy, and creativity. This gives raise to the need for new approaches in interdisciplinary team configurations. This is what is promised by Design Thinking. But how does Design Thinking delineate exactly from Requirements Engineering?

Design Thinking and Requirements Engineering: Two Distinct, Yet Complementary Approaches

With its growing relevance in agile software development, Design Thinking has gained recognition as a creative problem-solving method, particularly when the real-world problem is complex or "wicked" (Buchanan 1992). Industry studies have highlighted this significant development. For example, based on a survey of the Hasso-Plattner Institute (Schmiedgen et al. 2015), over 69% of Design Thinking practitioners and managers identified Design Thinking as one of the major contributors to conduct an efficient innovation process. In a survey of IBM by Forrester (2018), Design Thinking was reported to reduce development and testing time by 33%, equating cost savings of around $1.1 Mio per major software development project. Some researchers even consider Design Thinking a "modern form of requirements engineering" (Beyhl and Giese 2016, p. 288) addressing some of the aforementioned challenges in current Requirements Engineering practices. However, we argue that this is not the case. Design Thinking and Requirements Engineering emerge from different backgrounds and offer different tools and approaches aiming at different goals, even though these goals are complementary by nature, as explained next.

In principle and as elaborated in more detail in the next sections, when developing a software-intensive product, we need to accommodate essentially two perspectives. On the one hand, we need a profound understanding of the socio-technical and the operational context of the system under consideration. It is important to elaborate on what problems, needs, and goals stakeholders really have, and what the particularities of the domains including limiting (e.g., regulatory) constraints and demands are. This constitutes the difficult task of gaining a profound understanding of the too-often fuzzy goals, rather than requirements or solution proposals, what their implications are, and what possibilities these goals open for future products. On the other hand, we need to elaborate a solid foundation for the engineering of a software product where we clearly specify—as far as possible in a nontechnical, solution-independent fashion—what the elementary functional and nonfunctional properties of the software product are. Those properties build the basis for a variety of engineering and management activities ranging from architectural design over implementation and verification to project organization and planning activities such as effort and cost estimations.

The first perspective is what is typically in scope of Design Thinking, which describes a specific mindset and often nontechnical approaches to penetrate the problem space from a user perspective and to deliver nontechnical throwaway prototypes that allow to better understand that problem early on. The second perspective is what is typically in scope of Requirements Engineering which describes (engineering) methodologies, approaches, and tools to specify requirements in a detailed and testable way that facilitates subsequent development and management activities in a seamless manner. Here, capturing the problem domains and user perspectives is in many ways important (think, for example, in terms of UX), but not always central. A central task in Requirements Engineering is often to focus on operational environments and underlying infrastructures as well as their technical constraints, implications, and cost structures, but also on evidently demonstrating compliance to regulatory standards existing for many industries (think, for example, in terms of safety-critical systems or cyber-physical systems). In that sense, Requirements Engineering comes in many forms and interpretations which are all different from the principal ways of working in Design Thinking and yet they are all complementary to each other.

Design Thinking leverages interdisciplinary teamwork for a structured approach of ethnographic methods, and fast and simple (non-technical) prototyping cycles to produce innovative solutions in early product, service, and system development processes (Brown 2008; Kolko 2015). This rather diverging nature of problem solving is notably different from the more converging ways of Requirements Engineering practices in most software-intensive projects (Harte et al. 2017). The multifaceted opportunities of applying Design Thinking for Requirements Engineering are highlighted by several research community members. Vetterli et al. (2013) were one of the first who suggested bringing Design Thinking and Requirements Engineering together for developing software applications. Academics with a content-focused view (*what* value does Design Thinking add) have recognized its value in terms of product quality, user acceptance, and process speed, mostly in specific domains like learning environments (Soledade et al. 2013), social innovation (Newman et al. 2015), or health care (Harte et al. 2017). Academics with a more process-focused view (*how* does Design Thinking add value), examine usage schemes of Design Thinking with software engineering techniques and agile development toolkits. For instance, authors have investigated the integration of Design Thinking and Scrum (e.g., Häger et al. 2015; Przybilla et al. 2018) and found evidence for higher innovation potential stemming from a combination of both approaches. Although mainly practice-oriented literature suggests potential benefits of combining Design Thinking and Requirements Engineering, or more generally speaking Software Engineering, knowledge on how this could be done in a seamless manner remains still unclear (Beyhl and Giese 2016).

While Requirements Engineering is a rather mature discipline with a long-standing history in research and practice, resulting in a plethora of holistic methodologies, practices, and tools, there is still limited knowledge about Design Thinking as Yoo (2017, p.v) emphasizes in his call to "advance the intellectual foundation of Design Thinking" for Information Systems research. Little is known, in fact, about the specific impact on Requirements Engineering. A deeper understanding of Design Thinking would enable both communities, Requirements Engineering and Design Thinking, to evaluate its application purpose and potential for discovering and specifying requirements more thoroughly.

This is what is in scope of this chapter in the hope to provide a solid foundation for the remainder of this book.

Contribution and Outline

In this chapter, we elaborate on an effective integration of Design Thinking into Requirements Engineering. Note that we do not pretend that there would be one exclusive way of doing Design Thinking or Requirements Engineering. Rather, our aim is to introduce the mindset and common practices of both worlds, abstract from those practices by means of concentrating on the underlying outcomes (artifacts), and finally to use those resulting more simplified models for an integration of Design Thinking and Requirements Engineering which we further complement with practical experiences and recommendations. This provides a common basis for the various invited expert discussions captured in this book.

In the remainder of this chapter, we focus on the following contributions:

- First, we introduce the very fundamentals of Design Thinking and Requirements Engineering including the principles and practices as often found in literature.
- We then elaborate a first artifact model for Design Thinking that captures the essential concepts, approaches, and terms, and we will do the same for Requirements Engineering. We particularly concentrate on an artifact-centric view as a process-agnostic means that allows us to concentrate on the essential work products and their dependencies while abstracting from the particularities of surrounding, often very complex and unique specific-purpose processes.
- We use the artifact models for Design Thinking and Requirements Engineering to propose an integration of both.
- To use that integration not only as a conceptual foundation but also allow for its effective use in practice, we conclude by introducing different operationalization strategies on how to make efficient use of the introduced combination of Design Thinking and Requirements Engineering to create human-centered software-intensive systems.

Rather than merely focusing on presenting an academically oriented concept model, we aim at elaborating on essential terms, principles, and concepts while considering (and extending) the perspective on the practical relevance as many results emerge from academia-industry collaborations.

One central hope we associate with this introductory chapter is therefore to set the foundation for the subsequent invited chapters and to contribute to the ongoing debates and efforts in effectively integrating both worlds.

Previously Published Material

Note that the insights provided in this book chapter emerge from previously published material, among it the dissertation of the first author (Hehn 2020) as well as the long-term collaboration with the second author. In some parts, we will explicitly borrow from parts of the dissertation in a verbatim manner.

Conceptual Background

In the following, we introduce the background to the extent necessary for the contributions of this book chapter. We will first elaborate on the very fundamentals of Design Thinking before concluding with a brief introduction of Requirements Engineering.

Design Thinking as a Human-Centered Problem-Solving Approach

Design Thinking is referred to as "a human-centered approach to innovation that draws from the designer's toolkit to integrate the needs of people, the possibilities of technology, and the requirements for business success" (Brown 2012). The roots of Design Thinking date back to the late 1960s, when design academics examined the mental processes that underlie design activities and transformed them into normative guidelines for creative problem solving (Simon 1969). These studies have expanded the scope of design beyond the boundaries of product styling to a way of thinking that can now be universalized for a multitude of disciplines (e.g., management, business, software development, and engineering).

The paradigm of human-centered design is both starting point and foundation of all activities at all stages in Design Thinking (Brown 2008). Design Thinking solutions evolve from the triad of human values (desirability), technological feasibility, and business viability, combining expertise from the field of design, ethnologic and anthropologic research, engineering, and business economics. The dimension of desirability (what people want and need) anchors in deep empathy for users and is applied by involving relevant stakeholders systematically throughout the entire process. Diverse design techniques help facilitate the creative transformation of user knowledge and insights into new concepts. Subsequently, feasibility and viability are integrated and explored. The lens of feasibility (how technology can help), therefore, demands an exploration of organizational capabilities and technological options in order to translate the human-centered requirement into actual products and services. Assessing the third dimension of business viability (what is financially sustainable) entails evaluating market opportunities and their compliance with the business objectives of the organization. Given its integrative nature, Design Thinking can be applied to "all aspects of business and society" (Brown 2009, p. 3) and is equally relevant for designing tangible and intangible solutions, both in public and in private sectors.

Design Thinking on an Operational Level

On an operational level, Design Thinking is interpreted in three ways: as (1) a process with a sequence of steps according to a prescriptive process framework, (2) a toolbox with a collection of methods for situational support, and (3) a mindset with a set of human-centered principles to be internalized (see Fig. 2). While all three modes are interlinked, they result in different conceptualizations on a practical level. As Fraser (2011) suggests, "it takes a combination of the right mindset (being) and a rigorous methodology (doing) that unlocks a person's thinking, and that one must consider all three of these factors." (p. 71).

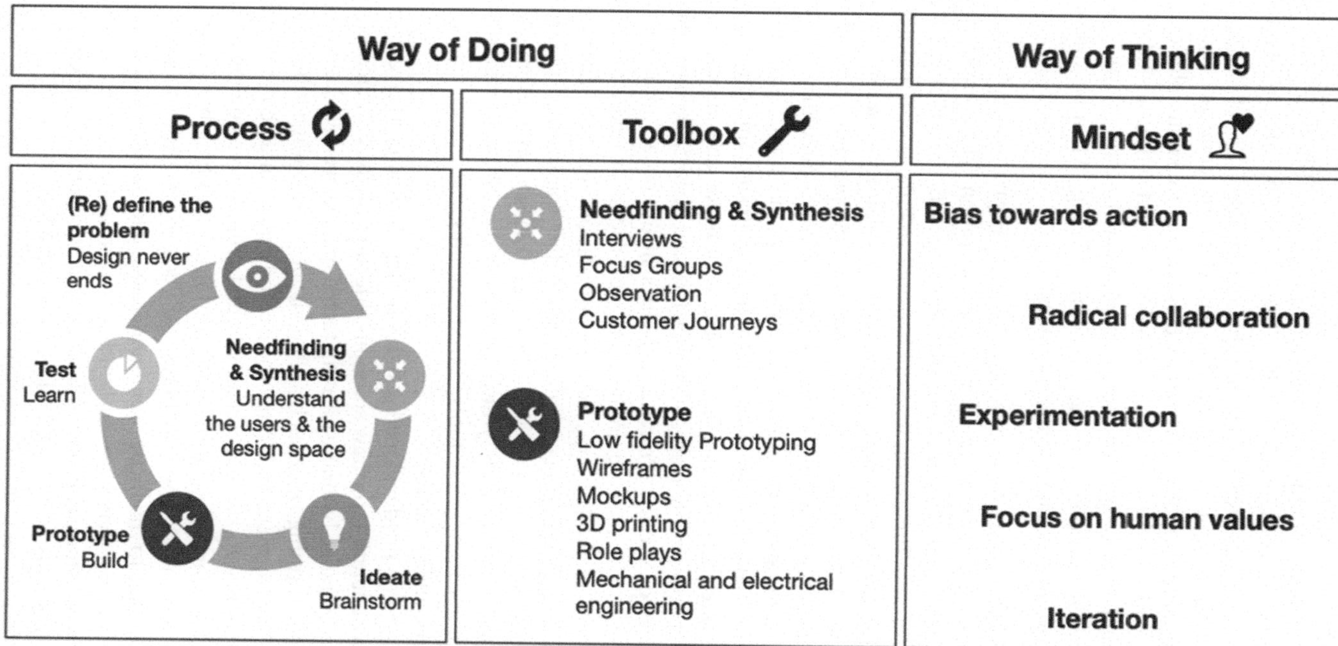

Fig. 2 Design Thinking as a process, toolbox, and mindset (see also Hehn 2020)

Design Thinking Process

Design Thinking process models, accompanied by a set of design tools, provide a supportive framework for practical use. Because of their specific character and clear instructions, those models are often utilized in Design Thinking education to provide a tangible, formalized approach to the Design Thinking concept. The normative Design Thinking process model is typically divided into two main phases: (1) problem exploration with problem definition, needfinding, and synthesis and (2) solution exploration with ideation, prototyping, and testing (ME 310 2010). The Design Thinking process as introduced in the left column in Fig. 2 can be summarized in five iterative steps as illustrated in the following:

1. *(Re-)Define*: The starting point of a Design Thinking process implies an intensive level of engagement with the topic under consideration. The complex problem is transformed into a single sentence (often starting with "How might we...?" or "What if...?") entailing a clear design challenge and, therefore, making the topic somewhat tangible. Activities in this stage are to identify sources of inspiration, assess relevant stakeholders and their impact on the problem, explore emerging trends and market adjacencies, and to prepare research directions.
2. *Needfinding and Synthesis*: In the second step the topic is concretized by collecting user data through field research. The design thinking team applies empathic research techniques to uncover hidden needs and unexpressed desires by finding out how people work, what they like and dislike, and how they interact with a product or service. Practical activities in the observation phase include interviews (e.g., with users, extreme-users, non-users, and experts) and a variety of observation methods (e.g., self-documentation, on-site observation, and shadowing). The acquired needfinding data is then transformed into meaningful insights about (unmet) user needs. Problem framing and reframing helps to identify patterns and ultimately develop a focus on where to create the highest value and impact. Applied tools are storytelling, scenarios, empathy maps, journey maps, and personas.
3. *Ideation*: Based on the developed insights in step 2, structured creativity methods support idea generation for new solutions. Ideation focuses on creating ideas and concepts (for instance by brainstorming techniques) as well as sketching them out quickly. Brainstorming rules such as "be visual," "encourage wild ideas," "defer judgment," "go for quantity," "stay focused on topic," and "build on the ideas of others" are applied to stimulate creativity and thinking outside the box.
4. *Prototyping*: Promising solution ideas are turned into tangible prototypes (e.g., (paper-) models, mock-ups, role-plays, storyboards, journey mapping, and short videos) in order to facilitate communication and feedback from end users. Therefore, it is not necessary to build perfectly well-engineered products, but rather simple versions and multiple alternatives in parallel, which focus on the most important aspects or highlight features for which feedback is crucial. Over the course of a project, prototypes usually evolve from so-called "Critical function/experience prototypes" (that define the core functionalities of the solution),

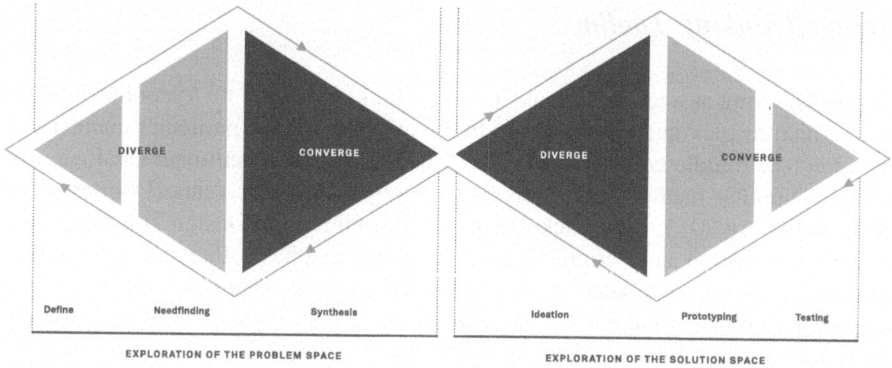

Fig. 3 Double diamond (see also Hehn et al. 2020, p. 26)

and "dark horse prototypes" (that challenge key assumptions and boundaries with visionary ideas) to "system prototypes" (that combine the most promising elements into one system vision) (Uebernickel et al. 2015).

5. *Testing*: The ultimate step is the collection of user feedback and definition of improvement opportunities. Since it is important to understand the physical interaction of the product in use, feedback from end-users and project stakeholders is processed for further concept enrichment and revision. Considering the new information, the Design Thinking team may then go back to earlier steps, often revising the point of view stage or even starting the entire process over again by doing additional research about a specific idea and its realization.

Another way of visualizing the innovation workflow is dividing the Design Thinking process into two exploration stages: (1) the exploration of the problem space and (2) the exploration of the solution space, both consisting of an interaction between information gathering (divergent activities) and information processing (convergent activities). This visualization is also called "Double Diamond" (see Fig. 3).

The problem space demands a diligent examination of the problem context by integrating all relevant stakeholders and the synthesis of all collected information to a clearly defined point of view, including needs and insights. The solution space encourages the generation of ideas and the creation of prototypes, which can be evaluated and tested with users. The process is repeated several times until a final solution can be presented. Reflection points are carried out during the process wherever necessary as they are crucial steps for adapting to novel information and developing deeper insights. Each cycle stimulates creativity and encourages rapid learning through trial and error.

Design Thinking Toolbox

Design Thinking as a toolbox breaks Design Thinking down into a set of techniques from which to pick and choose those that work best for the particular context and situation (see middle column in Fig. 2). A wide range of practitioner catalogues of Design Thinking methods and tools have emerged in recent years (Doorley et al. 2018; IDEO.org 2015; Uebernickel et al. 2015). In this case, Design Thinking is not so much considered a prescriptive process or a distinct phase of a process, but rather a bundle of handy and selective (design) methods and techniques for situational support. Examples of the most used methods that are attributed to Design Thinking are summarized in Table 1 (Hehn et al. 2018b).

Table 1 Examples of Design Thinking methods (adapted from Hehn et al. 2018b)

Method	Description	Phase
Stakeholder mapping	Analysis of all stakeholders that are affected by the design challenge	Define
Desk research	Desk research is known for collecting data based on literature and Internet research	Define
Framing and reframing	Framing and reframing is used to define the scope (and out of scope) of a project	Define
Interviewing	Conversation between two or more people where questions are asked by the interviewer	Needfinding
Observation	Observation and descriptions of happenings in the real world	Needfinding
Active listening	Technique to elicit needs by understanding and responding to what someone has said	Needfinding
Clustering	Technique to bundle ideas and statements into thematic buckets	Synthesis
Storytelling	Method for exchanging knowledge collected during needfinding	Synthesis
Insight formulation	Processes to distill and capture the most important learnings from needfinding	Synthesis
"How might we" questions	"How might we . . ." is a way of asking questions to initiate brainstorming but also entire projects	Ideation
Brainstorming	Brainstorming is a group creativity technique, mostly based on Osborn's method	Ideation
Brainwriting	A similar technique to brainstorming but all ideas are collected in written format before the information exchange within the group starts	Ideation
Paper prototype	Tangible representation of a product or service to facilitate testing	Prototyping
Role playing	Role playing is used to act out service scenarios quickly and simply	Prototyping
Sketches/ scribbles	Sketching and scribbling is all about drawing ideas and making them more tangible	Prototyping
Feedback capture grid	Framework to capture user, customer, or stakeholder feedback while testing (dimensions often are: Likes, Criticism, Ideas, and Questions)	Testing

Contrary to the process perspective, the toolbox offers an even more flexible way of using Design Thinking and tailoring it to specific project conditions. Thus, it can be integrated into the daily work routine and into existing company structures relatively quickly. However, since many of the Design Thinking techniques are not necessarily exclusive to this approach, it may raise the question from which point onwards to actually speak of Design Thinking.

Design Thinking Mindset

A growing number of authors stress that the core of Design Thinking goes beyond process models and tools (e.g., Kröper et al. 2010; Martin 2009). They perceive Design Thinking primarily as a mindset or general "design attitude" toward creative problem solving (see right column in Fig. 2). This entails the development of empathy, an open-minded and optimistic approach to generating insights and ideas, and the rationality to investigate and fit those ideas in compliance with the context. The main principles are highlighted in the following:

- *Design Thinking emphasizes human values as a starting point and foundation for all related activities* (Brown 2008). Understanding what people need and want anchors in a deep empathy for users and is achieved by systematically integrating a variety of stakeholder groups throughout the development process, both through direct dialog and non-obtrusive observation methods.
- *Design Thinking solutions are mainly generated through radical collaboration*, both with users and by composing a multidisciplinary project team that incorporates different functions and departments (Doorley et al. 2018). By encouraging interorganizational cooperation on the ground of common principles for a collaborative culture, Design Thinking can be regarded as a holistic framework for co-creation.
- *Design Thinking leverages abductive reasoning to constantly generate new information and consider alternative options early on.* The abductive nature of this way of working induces a "reflective conversation with the situation" (Schön 1984, p. 76) by looking beyond "what is" and exploring the logic of "what might be" to generate customer and business value (Martin 2009).
- *Design Thinking stresses a bias toward action.* This means that the preferred ways for gathering insights and feedback from stakeholders are hands-on activities such as experimenting with ideas, building prototypes, and testing them (Doorley et al. 2018).
- *Design Thinking is a fundamentally exploratory process that encourages rapid and iterative learning cycles.* According to the "fail early and often"-principle every iteration leads to further adjustments and new directions in the development process. In the long run, this iterative approach to development is supposed to mitigate risks of not meeting customer needs in the long run (Brown 2009).

Artifact-Based Requirements Engineering and the AMDiRE Approach

Similar to as it is the case for Design Thinking, Requirements Engineering, too, comes in various forms and interpretations while none is best for all purposes. In this chapter, we will not even try to introduce the discipline in its various interpretations to the extent they deserve, same as it is not our intention to promote any of the various (and often competing) approaches to Requirements Engineering. Rather, we aim at laying the foundation for Requirements Engineering that integrates the very Design Thinking tools and principles introduced above.

In principle, how Requirements Engineering is done in practice—including the artifacts created and the techniques used—depends on many factors such as surrounding software process models, application domains, industry sectors, and even engineering cultures including personal, subjective preferences in engineering teams. Those characteristics render Requirements Engineering approaches as something unique and barely standardizable with a one-size-fits-all solution. In response to this complexity in the choice of methods and approaches, various artifact-based approaches to Requirements Engineering have been elaborated over the last two decades. All those approaches capture the particularities of the envisioned domains and serve as reference models to guide the elaboration of precise requirements for those domains while offering the necessary flexibility in how to do it from the perspective of processes and activities. To this end, corresponding approaches offer blueprints of the results and their dependencies rather than a dictate for complex activities, tasks, or methods. This is what essentially reflects the artifact-centric philosophy. In such a philosophy, we concentrate on defining the artifacts, their contents, and their dependencies in a central model that constitutes the backbone of a (Requirements Engineering) project, and which leaves open when to create which artifact and which description technique to use (Mendez Fernandez et al. 2019). Such an artifact model then serves as a guide for engineers in elaborating their results (e.g., the specification of user requirements via use cases and capturing their relationship to acceptance test cases to support traceability) while leaving open the way how to do it (e.g., in an agile manner or a rather plan-driven manner).

In this book chapter, we rely on one specific artifact-based approach to Requirements Engineering which we use as integration point for Design Thinking. The approach we rely on is the Artifact Model for Domain-independent Requirements Engineering (short: AMDiRE). AMDiRE emerges as a concluding synthesis of the various approaches developed in recent years for different domains and with different partners from the relevant industries, e.g., Capgemini, Siemens, Bosch, BMW, or Cassidian. The AMDiRE approach thus emerged as the result of consolidating previously developed approaches and the lessons we learned during their development, evaluation, and dissemination.

Here, we focus on the very foundation of AMDiRE to the extent necessary in the context of our chapter. Details on the approach can be taken from previously published material (Mendez Fernandez and Penzenstadler 2014).

Overview of AMDiRE Components

Figure 4 illustrates all components included in the AMDiRE approach necessary to use it as a reference at project level. The central component of AMDiRE is defined by its artifact model. For the sake of simplicity, we see an artifact as a key deliverable of major interest that abstracts from contents of a specification document. It can be used as input, output, or even as an intermediate result in Requirements Engineering created along a particular task or method and by choosing a description technique (e.g., natural language, structured tables, figures, or models) as long as it complies with the artifact model as explained next. A more insightful introduction into what artifacts are in software engineering can be taken from our reflection provided in previous work (Mendez Fernandez et al. 2019).

For each artifact, we capture two essential views: a structure view and a content view. The structure view captures for each artifact type (e.g., "requirements specification") the content items to be considered (e.g., "use case model"). For each content item, we define the content view via the modelling concepts, e.g., the elements and content relations of a use case model and different description techniques that can be used to instantiate these concepts, such as an UML activity diagram. The structure model thus gives a simplified view on the content and is used to couple the contents to the elements necessary to define a process, e.g., roles, methods, and milestones relevant for a use case model. This is in scope of the integration of AMDiRE into company- and project-specific software process models (often referred to as static tailoring). The content model then guides by defining what is necessary to specify the content, e.g., scenarios, actions, and actors, which we use to create a use case model.

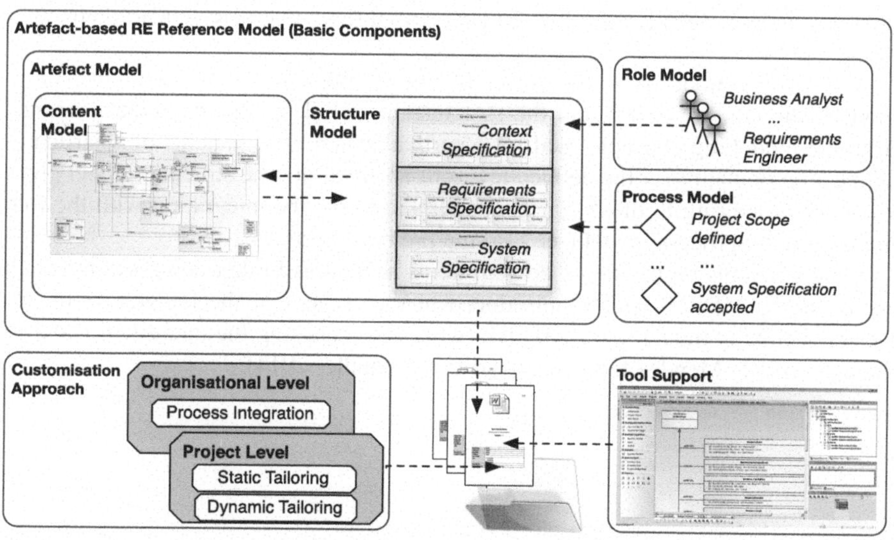

Fig. 4 Overview of AMDiRE components (see also Mendez Fernandez and Penzenstadler 2014)

Note that we consider—same as for activity-centric approaches to Requirements Engineering—elements of a process description, but instead of defining the process based on phases, activities, and methods, we define the process based on the artifacts to be created and their relationships, as well as related milestones for when these artifacts should be of sufficient quality to specify the next. Even though the content model supports the precision of the results in the flexible process definition, the process itself remains undefined. Regarding the methods and description techniques for creating the contents (e.g., UML or natural text), we leave open which one to choose, as long as the contents and relationships proposed by the artifact model are specified.

AMDiRE Artifact Model

The AMDiRE artifact model comprehends concepts used to specify the contents of the artifacts over three levels of abstraction: the Context Layer, the Requirements Layer, and the System Layer (Fig. 5). Each of those levels of abstraction features a specified number of content items that are detailed in concepts used for a stepwise refinement of the various (modelling) views we have on a system. The context layer considers the context of a system, i.e., the domain in which to integrate the system such as the business domain with the business processes to be carried out. The requirements layer considers the system from a black-box perspective. That means, we specify the requirements on the system and the user-visible functionality from a perspective in which the system is intended to be used, without giving details about its technical, internal realization. That view is captured by the system layer which provides the glass-box perspective on the internal (logical and technical) realization of the system.

The artifact model is in the center of our attention and consists of two basic models: the content model and the structure model. The content model abstracts from the modelling concepts used for a particular family of systems in a particular application domain over the defined levels of abstraction. The structure model gives a logical structuring to those concepts and is used for the integration with the role model and the process model (see also the previous section).

Finally, details on the single content items as well as further components which accompany AMDiRE will be introduced in the context of the integration of our Design Thinking model into AMDiRE (while also referring the interested reader to the main article Mendez Fernandez and Penzenstadler 2014).

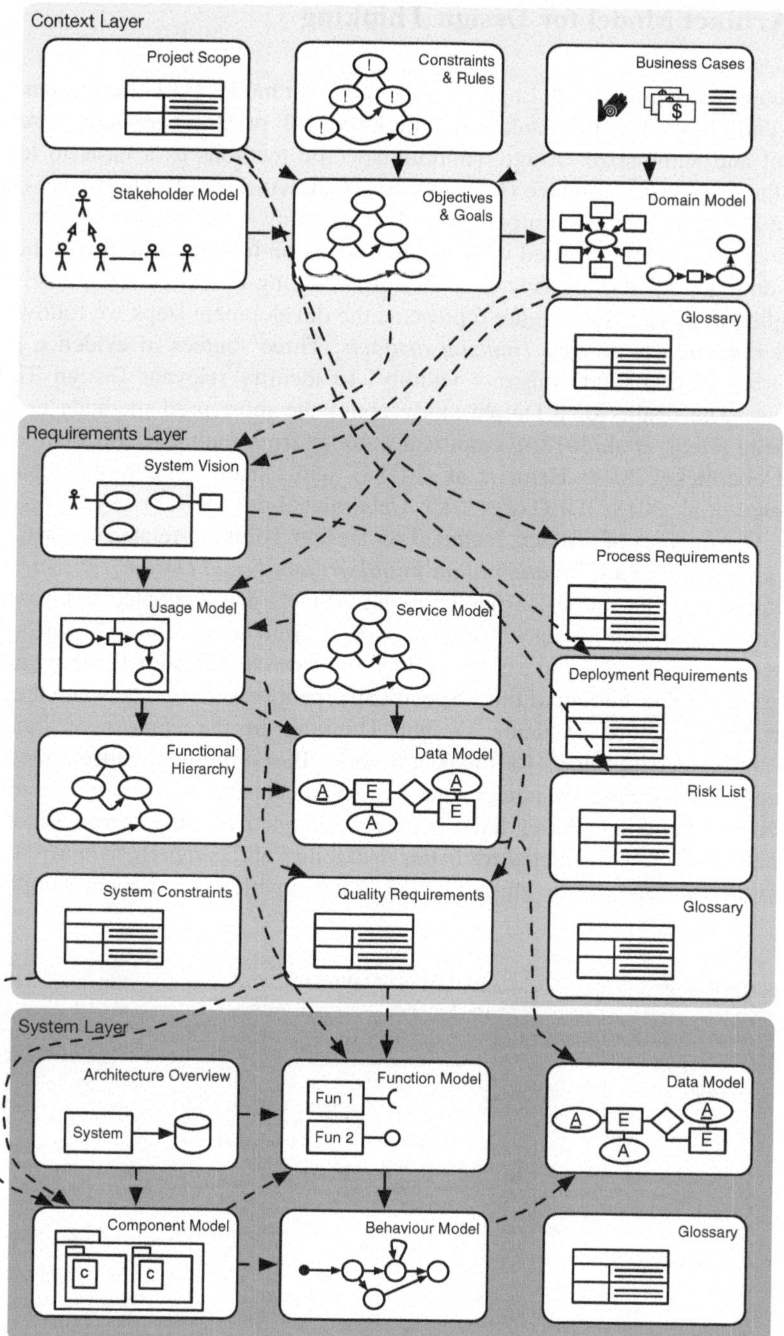

Fig. 5 AMDiRE artifact model (simplified view on structure model, see also Mendez Fernandez and Penzenstadler 2014)

An Artifact Model for Design Thinking

In contrast to Requirements Engineering, no artifact model exists for Design Thinking—until now. We have taken the multitude of practitioners' compendia that present and summarize Design Thinking-specific methods as a basis to logically infer the results they produce (i.e., artifacts) (Gutzwiller 1994). Hence, we can rely on the available literature corpus as well as the knowledge we accumulated in our own more practically oriented work as the foundation for determining, synthesizing, and summarizing the artifacts in a Design Thinking-based artifact model that is described in this section. Figure 6 presents the development steps we followed.

Identification of Design Thinking artifacts: Three sources of evidence provide data triangulation (and construct validity) to identify relevant Design Thinking artifacts. The results of a Delphi study about the most used methods in Design Thinking (Hehn et al. 2018b), empirical findings from multiple case studies (Hehn and Uebernickel 2018; Hehn et al. 2018a), and existing practitioner catalogues (Doorley et al. 2018; IDEO.org 2015; Uebernickel et al. 2015) serve as our main basis. The final set of artifacts included 65 Design Thinking-related artifacts.

Construction and evaluation of an initial artifact-based Design Thinking model: The initial model with 65 Design Thinking artifacts was evaluated in unstructured interviews with four Design Thinking experts from academia and industry. The experts were required to have either applied or researched Design Thinking methods for a considerable amount of time. Specifically, people were chosen when they had a proven track record of using Design Thinking in the context of innovative software-intensive projects for the past 3 years. Based on the feedback, three main findings evolved: First, the completeness of relevant artifacts and their attributions to the Design Thinking phases have been corroborated by all experts. Second, the original structure was adapted for better readability and comprehensibility from top to bottom according to the chronological order in which they typically appear in a

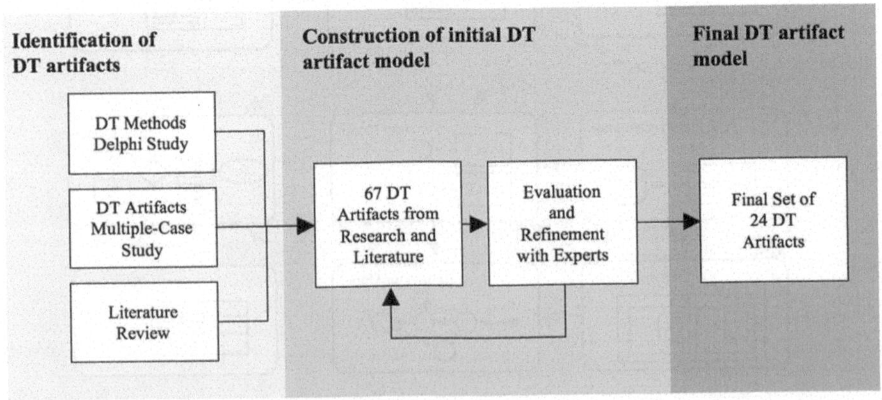

Fig. 6 Development steps of a Design Thinking-based artifact model (see also Hehn 2020)

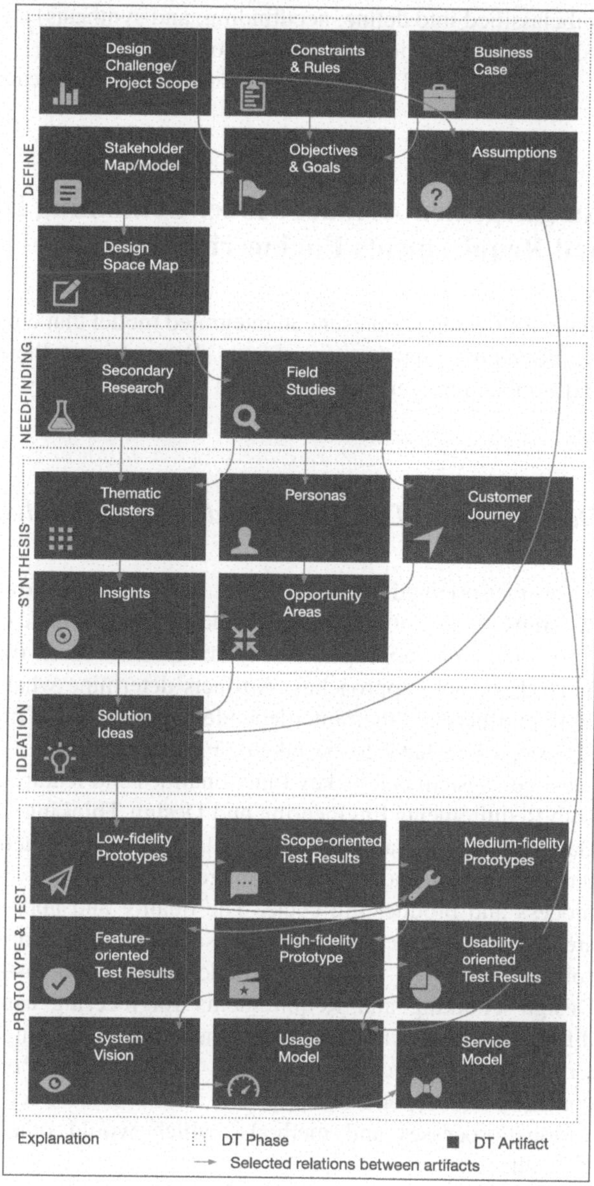

Fig. 7 Design Thinking artifact model (see also Hehn 2020)

project. Third, the model was refined to fit the frame of reference in terms of granularity of the artifacts. The second version of the model encompasses 21 artifacts and is presented in this book chapter.

Construction of the final artifact-based Design Thinking model: The revised and final version of the artifact-based Design Thinking model is visualized in Fig. 7. It encompasses 24 Design Thinking artifacts structured into problem-oriented artifacts (subclassified into define, needfinding, and synthesis) and solution-oriented artifacts (subclassified into ideation and prototype and test).

A more detailed description of each content item can be found in the Appendix.

An Integrated Artifact Model Combining Design Thinking and Requirements Engineering

In the following, we present an integrated model that combines Design Thinking and Requirements Engineering artifacts. We motivate the development, the structure, and implications for researchers and practitioners.

Development of an Integrated Artifact Model

An artifact-oriented reference model, such as those shown in the sections before, and that aims at integrating Design Thinking into a holistic engineering context is, as we argue, the only appropriate way to accommodate the variety of processes and methods of both approaches. Artifacts determine what must be accomplished (the work products and their interdependencies) instead of how it has to be accomplished (the steps that have to be taken). Further, defining a comprehensive view of the "desired" system and its key functionalities and features is an important objective of both Requirements Engineering and Design Thinking. The artifacts produced along Design Thinking and Requirements Engineering activities are used to support product design and project management decisions throughout the development process and product life cycle. The quality and appropriateness of these artifacts are therefore imperative for the successful development and acceptance of a software-intensive system. A model that encompasses the relevant artifacts of Design Thinking and Requirements Engineering can outline the synergies and differences between both approaches. While keeping a consistent structure and terminology, this condensed view focuses on the created work products, their contents, and dependencies, and it allows to abstract from their particularities of various processes and methods, which would otherwise render a comparison difficult.

Our integrated artifact model, therefore, contains and structures all the artifacts referenced, modified, or created in Requirements Engineering and Design Thinking in software-intensive development projects. To be useful, the model should support the re-use of knowledge and should be tailorable to certain situations in an efficient manner. The aim is to integrate Design Thinking and Requirements Engineering artifacts to simplify the adoption and configuration (i.e., usage schemes) of Design Thinking for Requirements Engineering.

Our goal was to establish a reference model that should:

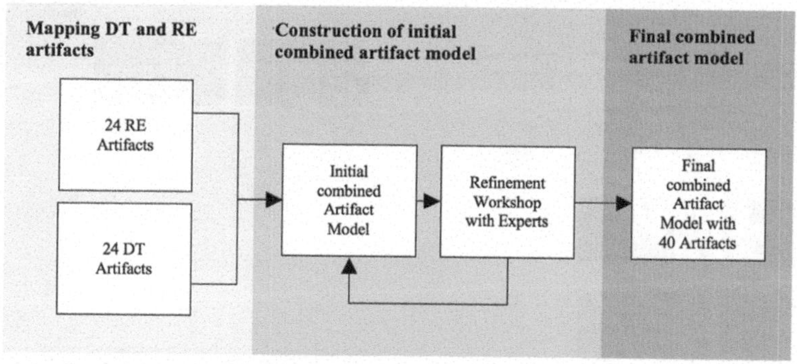

Fig. 8 Construction and evaluation of an integrated artifact model (see also Hehn 2020)

1. Support the integration of both approaches respecting their different "flavors."
2. Provide flexibility in the way of working to cope with the various influences in individual project environments and for organizational needs.
3. Enable a reproducible creation of work products in the context of innovative software-intensive development projects.

Similar as done for the development of the artifact model for Design Thinking itself, we show the steps for the construction and evaluation of our final combined artifact-based reference model for Design Thinking and Requirements Engineering in Fig. 8.

The process of mapping artifacts from Design Thinking and Requirements Engineering was performed by two experts in Design Thinking and Requirements Engineering. The comparison was performed with 24 Design Thinking artifacts and 24 Requirements Engineering artifacts. Based on these activities an initial integrated artifact model for Design Thinking and Requirements Engineering was created. This model has been tested with Design Thinking and Requirements Engineering academics and practitioners to adapt the relevant artifacts and their interdependencies for a comprehensive overview. Details on the methodological approach can be taken from Hehn (2020).

Integrated Artifact Model

The integrated artifact model is presented in Fig. 9. It establishes a blueprint of relevant artifacts, i.e., the work results, contents, and dependencies of Design Thinking and Requirements Engineering. All artifacts are denoted in rectangles including the name of the artifact and a number. Associations depict relations between the artifacts, however not exhaustively, for reasons of reducing visual complexity. The Design Thinking phases (dotted line) provide a substructure for organizing the Design Thinking artifacts.

Table 2 summarizes the elements used to compose the artifact model.

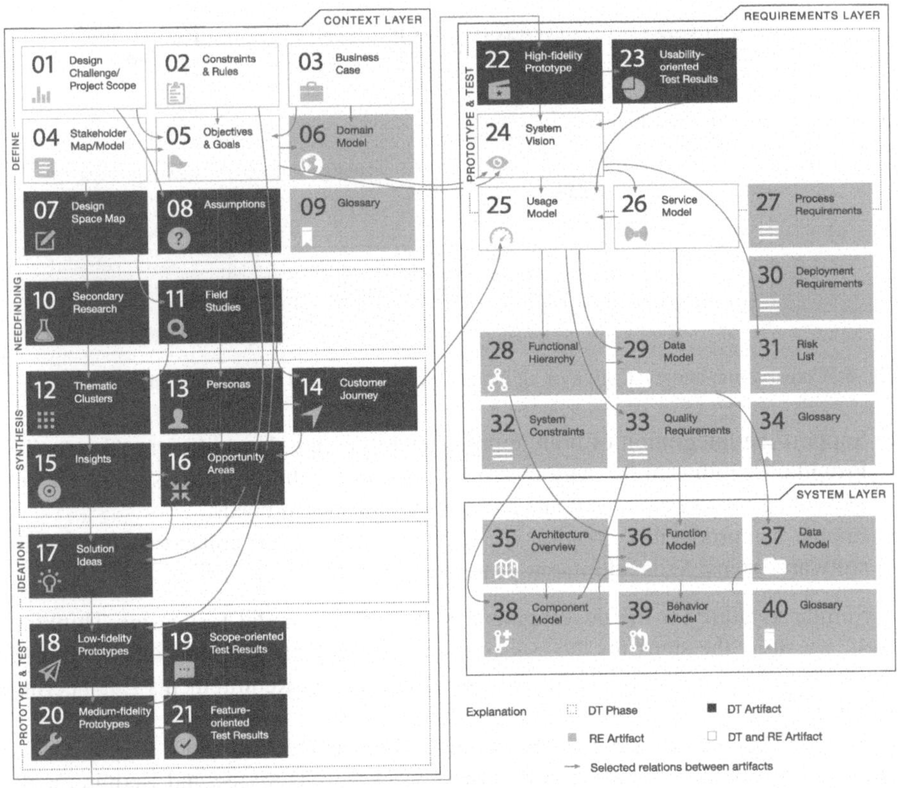

Fig. 9 Integrated artifact model (see also Hehn et al. 2020, p. 27)

Table 2 Overview of elements in the integrated artifact model

Representation	Description
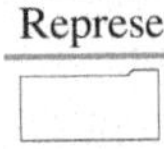	The folder box denotes the layers context, requirements, and system as the overarching structure of the artifact model
	The dotted line indicates the Design Thinking phases (Define, Needfinding, Synthesis, Ideate, Prototype, Test) for means of comprehensibility
	The dark rectangle denotes a Design Thinking artifact including the artifact name, a number in the artifact model, and an icon
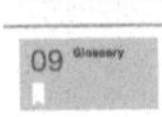	The gray rectangle denotes an requirements engineering artifact including the artifact name, a number in the artifact model, and an icon
	The white rectangle denotes a combined artifact (design thinking and requirements engineering artifact) including the artifact name, a number in the artifact model, and an icon
	The arrow denotes a unidimensional relation between artifacts. It expresses an input–output relationship

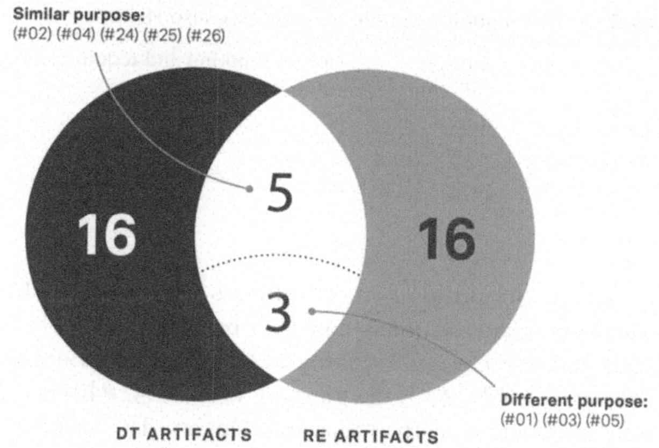

Fig. 10 Distribution according to artifact type (see also Hehn 2020)

The overall structure of the model is orientated along the three layers of the AMDiRE model (context, requirements, system)—each capturing a collection of relevant content items from Design Thinking and/or Requirements Engineering.

As discussed earlier, the *context layer* covers the information relevant to defining the context and includes, for example, the overall project scope, stakeholder information, a domain model, and assumptions of the project team, and underlying goals and constraints. Hence, much of the information captured in Design Thinking concentrates on this layer.

The *requirements layer* encompasses what is necessary to operate in this context and captures, for example, the system vision, high-fidelity prototypes, a usage and behavior model, and the function hierarchy as entry points for the system layer. Similar to the context layer, much of the information here is documented using natural language, occasionally reflected, however, also in models (e.g., for data and functional perspectives on user-visible system behavior).

Finally, the *system layer* includes information on how the system is to be realized and includes, for example, a logical component architecture and a specification of the desired behavior, e.g., via function models. Again, information within this layer is documented using both, natural language and conceptual models (data, function, behavior).

The integrated artifact model consists of three artifact types that encompass 40 content items with various relations. Out of all content items, 16 can be associated with Design Thinking, 16 with Requirements Engineering, and eight with both (see Fig. 10). The latter can be further distinguished into artifacts with similar semantics but different purposes (three out of eight). These include the design challenge/project scope (#01), the business case (#03), and the objectives and goals (#05). The main reason for their different purpose is that in Requirements Engineering these artifacts have a convergent nature while in Design Thinking they can be considered as open because they provide the opportunity for a broad context exploration.

The distribution of artifact types according to the specific layers in the artifact model is depicted in Table 3.

Table 3 Distribution according to layer (see also Hehn 2020)

Layer	Design thinking	Design thinking and requirements engineering	Requirements engineering	Total
Context	14	5	2	21
Requirements	2	3	8	13
System	0	0	6	6

The model positions most artifacts within the context layer (21). Most Design Thinking-related artifacts can also be found here (14 Design Thinking only artifacts and five Design Thinking & Requirements Engineering artifacts). Next to the data model (#29, #37) the glossary (#09, #34, #40) is a Requirements Engineering-only artifact that can be found in all layers. This artifact type is revised based on the specific layer objectives. Starting in the context layer, the design challenge/project scope (#01) defines the relevant problem and primary scope of a project. Within this realm, the stakeholder map/model (#04) captures the most relevant stakeholders and their relationships. They provide one important rationale for the requirements and goals of the system (#05). The domain model (#06) contains context information and constraints (#02) about the operational environment connecting it to the requirements layer. Design Thinking artifacts complement and expand these mainly Requirements Engineering-related artifacts with a broad and human-centered perspective. For example, field study results (#11) and insights (#15) help to frame the project scope (#01) and inform specific use cases and scenarios (#25, #26) as defined in the requirements layer. Low- and medium-fidelity prototypes (#18, #20) are mainly leveraged to better understand stakeholder needs and system context.

The requirements layer contains five Design Thinking-related artifacts (two Design Thinking only artifacts and three Design Thinking and Requirements Engineering artifacts) and eight Requirements Engineering artifacts. The system vision (#24) denotes the general concept and idea of the intended system. High-fidelity prototypes (#22) are a way to visually enrich the system vision (#24) and to illustrate the key functionalities and general form of interaction (app, desktop solution, etc.). Agreed upon by the relevant stakeholders, a system scope, i.e., major features and use cases as well as its constraints (#32), is specified. A service model (#26) defines the services the system shall offer complementary to the use cases defined through a use case model (#25). User-visible system functions are structured in a functional hierarchy (#28), which is the entry point into the system layer.

The system layer holds six Requirements Engineering artifacts and none of them are related to Design Thinking. While the context and the requirements layers include the information aspects that are typically found in Design Thinking- and Requirements Engineering-related artifacts, the system layer includes the items addressing what is known as the solution space and providing the interface for Requirements Engineering into design activities. In the system layer, the functions of the functional hierarchy (#25) are related to components (#38), a functional model (#36), and their internal behavior (#39), which also provides the basis to identify the data model (#37).

A more detailed description of each content item can be found in the Appendix.

Organizational Model

The integrated artifact model can be seen as a foundation for a more comprehensive organizational model that includes the following components: (1) the artifact model specifies what needs to be produced or exchanged; (2) the role model describes who should produce it and which particular responsibilities are needed; (3) the activity model describes what to do in order to create, modify, or use an artifact; (4) the process model denotes when the artifacts, roles, and activities should be produced or performed; and (5) standards and tools conceptualize with what all of the above-mentioned activities are performed (Mendez Fernandez and Penzenstadler 2014).

Figure 11 shows the artifact types in relation to roles and responsibilities (left side) and in relation to milestones (right side), which can be used to integrate the model into a process. We distinguish the Design Thinking- and the Requirements Engineering-view.

Note that in Requirements Engineering and in accordance with AMDiRE, we assign one role for each artifact type. Each role has the responsibility independent of other potentially supporting roles such as those provided by the surrounding software process model (e.g., product manager), and independent of whether same people are assigned to different roles in a project. The Business Analyst has the responsibility for the context specification, the Requirements Engineer has the responsibility for the requirements specification serving also as a mediator between the business analyst and the system architect. That system architect, finally, has the responsibility for the system specification. In Design Thinking, a multidisciplinary

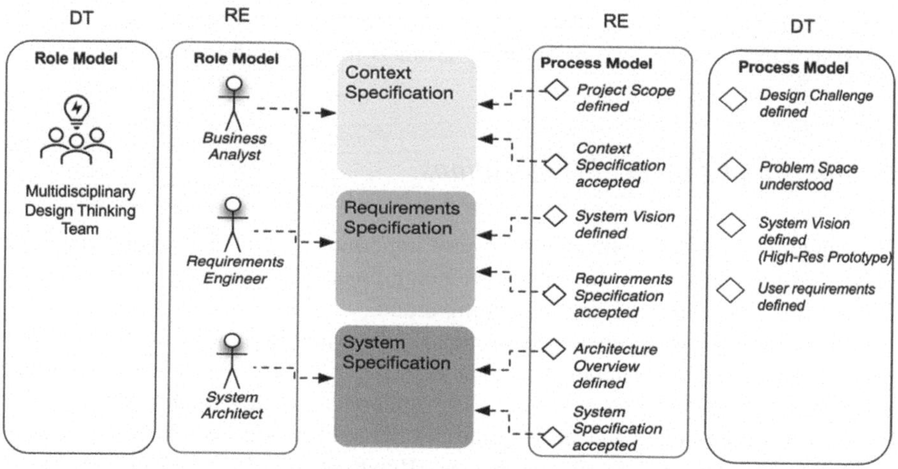

Fig. 11 Overview of artifact types, roles, and milestones

team takes up the role to define the context and system vision. Often this team is drawn from various disciplines to integrate diverse perspectives constituting an important aspect in stimulating creativity and generating the potential for more comprehensive and original results. The willingness to cooperate with different people is an important aspect in Design Thinking practice since solutions are mainly generated through collaboration, both with users and by composing a multidisciplinary project team (around six team members). Typically, Design Thinking team structures are not subject to hierarchies and departmentalization but rather a way of radical collaboration that allows leadership to pass in-between members. Team members drawn from various disciplines integrate diverse perspectives constituting an important aspect in stimulating creativity and generating the potential for more comprehensive and original results. The versatile Design Thinker has acquired the position of a general problem solver possessing strengths in two dimensions which are commonly visualized as a "T-shape." Deep Knowledge corresponds to the academic expertise or a depth of skill that allows the Design Thinker to adapt their knowledge to the problem and make tangible contributions to the result. Broad knowledge and skills represent the ability to reach out to other specialists coming from a wide range of disciplines entailing a general openness to new ideas, people, and ways of doing.

For each artifact type, we furthermore define two milestones: An entry-level milestone indicates the point in time in which the first content item is expected to have a sufficient maturity in its content; for instance, the system vision in the requirements specification comprises an overview of the major use cases; its definition and agreement indicate that the use cases are succinctly defined to be further refined and modelled and, thus, allowing, for example, for first cost estimations based on function points. The second one indicates when the corresponding artifact is formally accepted.

Those milestones are sufficient for process integration and instantiation as they give us the opportunity to formally embed the artifacts into project-specific decisions. Therefore, we enrich those existing milestones in analogy to the AMDiRE milestones to cover the Design Thinking artifacts following the same logic.

Findings and Practical Implications

Our integrated artifact model offers several important insights and implications for using Design Thinking in the context of Requirements Engineering. In the following, we highlight those we deem most important.

Various commonalities between Design Thinking and Requirements Engineering can be seen if the latter is understood as an iterative approach. The differences should be seen as complementary activities. The integrated artifact model distinguishes between more problem- and more solution-oriented artifacts, which addresses the principles of both Design Thinking and Requirements Engineering. Problem-oriented artifacts contain information about the underlying problem context including the goals and needs of stakeholders as well as specific system conditions or constraints. Solution-oriented artifacts contain information about the corresponding system vision and how to solve the problem stated in the project description.

The integrated model shows that Design Thinking mainly contributes to early Requirements Engineering activities with up to 14 additional context artifacts for a comprehensive understanding of the problem domain. Accordingly, Design Thinking expands the toolbox for Requirements Engineering by emphasizing the creation of artifacts that describe the relevance of the system vision. Design Thinking could even be exclusively used to perform these activities. A complementary approach of Design Thinking and Requirements Engineering, however, seems necessary for shaping the requirements layer. While both concepts produce overlapping artifacts (system vision, functional requirements, usage, and service models), their realization might take different forms. Design Thinking uses mainly a high-fidelity prototype to describe the system vision and key functionalities. Requirements Engineering specifies the same mainly by using rich pictures and class diagrams. In addition, other requirement types, such as quality or deployment requirements are predominantly specified with common Requirements Engineering techniques. Requirements Engineering is exclusively used to specify system artifacts and to provide the interface to system design activities. Hence, Requirements Engineering also expands the toolbox of Design Thinking.

Following our AMDiRE role model as described in Mendez Fernandez and Penzenstadler (2014) (see Fig. 11), implications can be seen in expanding the knowledge of business analysts with Design Thinking skills and, vice versa, in equipping design thinkers with Requirements Engineering skills to gain appreciation for subsequent software design activities. Lauenroth (2018) calls this role "digital designer" and defines them as "someone who is capable of creating a vision for digital products, processes, services, business models, or even entire systems, free from technical or organizational obstacles as well as apparent reservations (outside-in thinking). Digital designers are also capable of ultimately turning this vision into reality. They transfer (technological) possibilities into (new) product/process/service/business model/system design. To do all of this, digital designers must be skilled in design and the available technologies and be capable of interacting with all stakeholders" (p. 8). For training providers, the integrated artifact model can support the development of new training programs and learning formats about combining Design Thinking and Requirements Engineering. A new role with skills and talents in both approaches may be fostered. Current training courses in Design Thinking or Requirements Engineering can be enhanced by integrating the respective other approaches to gain an understanding of the benefits and shortcomings of the two incorporated concepts.

For project managers, several contributions can be seen. First, the model can be considered a support system to define and distinguish responsibilities in a project. Project roles can be directly coupled to the creation of artifacts, for which they must take responsibility. Second, project managers can assign completion levels and establish progress control for the creation of artifacts. Quality assurance metrics can help to objectively measure the degree of completeness of an artifact in the artifact-based reference model. Third, the model ensures flexibility for integrating processes and customizing the reference model at the project level. The combined model allows for variations of the created artifacts in response to individual project characteristics. For example, by defining the content focus of the project, the creation of either Design Thinking or Requirements Engineering artifacts might be of greater help as each approach emphasizes a different content type. For example, to better understand the user and business context, the creation of Design Thinking artifacts might be preferred. Requirements Engineering artifacts should be at the center of attention to better describe the technical perspective and answer feasibility questions. Teams may also jump back and forth between both approaches if new questions come up in one or the other area. Fourth, the model can act as a basis for effective requirements management, where the objective is to administrate the outcome of Requirements Engineering activities. This administration includes, for example, progress and traceability control, impact analyses, or risk mitigation (Jönsson and Lindvall 2005). A structured and consistent content specification is a prerequisite to perform such activities. Hence, the integrated artifact model can enhance the effectiveness of requirements management activities due to its defined set of interdependencies and chosen artifacts.

For team members of software-intensive projects (i.e., requirements engineers, business analysts, or design thinkers) the model offers a blueprint for creating syntactically consistent and complete results with respect to the respective application domain. While not all artifacts from the model must be considered in every project, the overview still serves as an orientation and connection to further design and development activities. The latter point is especially of interest for Design Thinking as this has been continuously criticized to be insufficiently linked to development processes (e.g., Häger et al. 2015).

Operationalization Strategies

In the following chapter, we present three operationalization strategies to integrate Design Thinking into Requirements Engineering when designing innovative software-intensive systems.

Overview

The integrated artifact model enables a flexible creation of the introduced Design Thinking and Requirements Engineering artifacts. This means that the decision on which and when artifacts should be produced needs to be customized according to specific project characteristics. To provide a guideline three operationalization strategies are proposed to integrate Design Thinking and Requirements Engineering in different ways. The strategies reflect existing research findings about integrating

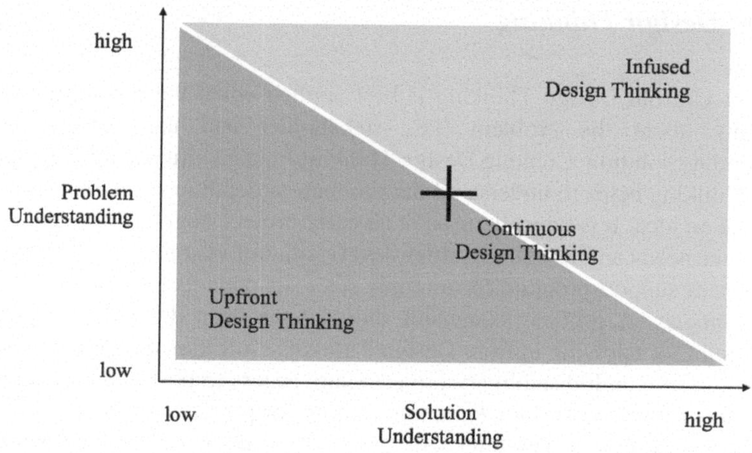

Fig. 12 Navigating upfront, infused, and continuous Design Thinking strategies (see also Hehn 2020)

Design Thinking into software development practices (e.g., Dobrigkeit and de Paula 2019; Lindberg et al. 2012; Hehn and Uebernickel 2018).

We suggest the following three strategies: (1) Run Design Thinking prior to applying Requirements Engineering practices (upfront Design Thinking); (2) infuse the existing Requirements Engineering process ad hoc with selected Design Thinking tools and artifacts (infused Design Thinking); or (3) combine the previous two strategies and integrate Design Thinking into Requirements Engineering practices on an ongoing basis (continuous Design Thinking). The ratio between Design Thinking and Requirements Engineering differs within the three proposed operationalization strategies. The better the original problem is understood, the more activities are biased toward straightforward design and implementation tasks (i.e., Requirements Engineering artifacts) (see Fig. 12). The less it is understood, the more activities are directed toward context understanding and problem exploration (i.e., Design Thinking artifacts). Thus, the defined project objective and context are the guiding parameters for the selection of an appropriate operationalization strategy.

Three Strategies to Operationalize and Integrate Design Thinking

In the following, we introduce our three strategies to operationalize our integrated Design Thinking approach. For each, we follow a structured approach of listing objectives, prerequisites, key activities, necessary roles, and outcomes followed by showing an exemplary practical case. This shall make our strategies more tangible.

Upfront Design Thinking

Objective: Upfront Design Thinking is best applied when there is a high level of uncertainty about the problem (i.e., stakeholder and user needs) and the corresponding solution. Creating Design Thinking-related artifacts through applying Design Thinking helps to understand the problem in depth and to define the overall concept of an idea. It is typically used at an early project stage to provide clarity for unclear user needs and to define a (high-level) solution vision (e.g., "How does the future patient support program for multiple sclerosis patients look like?").

Prerequisites: A problem statement should have been defined as a minimum starting point for applying upfront Design Thinking. Additional required conditions are the setup of a multidisciplinary project team, access to potential users and other stakeholders as well as Design Thinking training for project members.

Key activities: Design Thinking activities are typically performed in the form of a pre-project to identify relevant features that are worth implementing. The Design Thinking process model (define, needfinding, synthesis, ideation, prototyping, testing) guides through a cyclical creation of context and requirements artifacts. The outcome is used as a basis for performing further Requirements Engineering activities that complement Design Thinking artifacts with Requirements Engineering specific ones.

Roles: Two roles during the upfront mode are required. First, the Design Thinking team is responsible for planning and executing the activities. This team consists of four to six people from different areas of expertise depending on what knowledge will be relevant for the project, including, for example, subject matter experts, IT, marketing, sales, design personnel (Häger et al. 2015). Second, a person or group of people, who has defined the initial design challenge and project scope, is defined as the project sponsor. The person in this role typically provides continuous feedback to the team and connects it with others to enable synergistic effects and avoid duplicate efforts (Häger et al. 2015). The following two roles are optional: First, an extended team of (internal) experts that provide further domain knowledge and expertise for the Design Thinking team. Second, a Design Thinking coach or coaches who support the project team with methodological guidance. They introduce Design Thinking techniques, facilitate team meetings, and ensure that the team is focused on delivering the tasks and artifacts. As such, the coach should have a profound understanding of Design Thinking to provide useful techniques and guidance at appropriate times (Häger et al. 2015).

Outcome: The main deliverable of the upfront strategy is a clear system vision as a basis for performing further Requirements Engineering activities. The system vision usually takes the form of a mockup (i.e., high-fidelity prototype). Along the way, the team will create a comprehensive set of Design Thinking artifacts, which should make it clear why each aspect of the prototype is intended in the way it is designed. High-level user stories and a list of usability requirements based upon test results accompany the set of artifacts created by following the Design Thinking process.

Case Example
The international Alpha Insurance company wanted to develop a new service for their new target group of "young professionals." A project team stemming from five different business functions (marketing, IT, actuary, product manager, claims) spent 40% of their time to follow the Design Thinking process in an iterative manner for 3 months. The solution vision resulted in a tested medium-fidelity prototype for a digital on-demand insurance that could be activated and deactivated based on the user's preferences. The Design Thinking team handed over the prototype to the implementation team for further specification, testing, development, and market introduction. Transferred artifacts included a project documentation with 20 field studies, two personas, five opportunity areas, and six low-fidelity prototypes with learnings about failures. The final solution vision (in form of a mockup) specified key features and their usability. The implementation team performed tests to validate these features, their usability, and their service model.[4]

Infused Design Thinking

Objective: The main goal of this strategy is to support existing Requirements Engineering activities with selected Design Thinking techniques. This includes, for example, activities to clarify fuzzy requirements, foster creativity, gain new ideas, or to better understand user needs.

Prerequisites: The prerequisites for applying this strategy depend on the specific problem to be addressed. The problem should have a clear scope. The prerequisites as described in the previous still apply.

Key activities: An infused approach makes use of selected artifacts and leverages selected methods from the Design Thinking toolbox and integrates them into an existing Requirements Engineering process. In case of challenges encountered during the Requirements Engineering process, Design Thinking tools can be initiated; hence, their application is ad hoc. The main activity of this strategy is the setup of focused workshops with selected Design Thinking tools (Dobrigkeit et al. 2018). These workshops can last 3 h or several days depending on the objectives. For example, the goal of a workshop to generate new solution ideas could be formulated like this: "Create ideas to optimize the user interface of our platform, making it look and feel more emotional, and letting it appear less technical." This session used persona and customer journey artifacts to brainstorm new ideas.

Roles: In the infused setting, the people or person performing the Requirements Engineering activities are the addressees of receiving Design Thinking guidance in the form of workshops. Other workshop participants with different areas of expertise

[4]This case has also been published in Hehn et al. 2020 and Hehn 2020

may be added, e.g., subject matter experts, IT, marketing, sales, design, depending on what knowledge will be relevant to achieve the workshop goal. A workshop typically consists of 5–20 participants. Like the upfront approach, a Design Thinking coach introduces the selected Design Thinking techniques and moderates the workshop and team discussions. The project sponsor can also be integrated to provide feedback and define the context for the general direction of the workshop.

Outcome: Due to the flexible approach of the infused strategy, the outcome is situation-dependent based on the previously defined objectives. The deliverables can be (new) features, user requirements, or test feedback—all following the Requirements Engineering process. In the context of the combined artifact model, this means that the creation of Requirements Engineering artifacts is enhanced by a selected set of Design Thinking artifacts.

Case Example
Beta Enterprises is an international electronics group that wanted to evaluate the possibilities of smartphone applications (e.g., emergency apps, task lists, and maintenance procedures) for container ships in a marine context. The main goal was to define requirements from a user point of view and to foster creativity for solution finding. In a highly regulated environment, a Design Thinking infusion was chosen to support the ongoing Requirements Engineering activities with selected tools from needfinding and prototyping. Five Design Thinking infusion sessions (1–2 days) were conducted within 5 months. Produced artifacts included field studies for precise user requirements (it was the first time the team had been in close contact with marine captains) and tested medium-fidelity prototypes to strengthen service and usage models. According to the workshop participants, having direct user contact raised the confidence level in the success of the intended solution. Initial concerns about not finding interview partners in a highly sensitive B2B setting turned out as unjustified.[5]

Continuous Design Thinking

Objective: The main goal of this strategy is to integrate Design Thinking principles with Requirements Engineering activities on a continuous basis. Beyond the specific project context, this can also become part of an organizational change program or corporate strategy.

Pre-requisites: Continuous Design Thinking is recommended when addressing complex ("wicked") problem settings, which require continuous user involvement along all software engineering activities. In addition to the prerequisites described for the previous two strategies, (selected) project members should possess both Design Thinking and Requirements Engineering knowledge.

[5] This case has also been published in Hehn et al. 2020 and Hehn 2020.

Key activities: Continuous Design Thinking utilizes the Design Thinking mindset as guiding principle. On an operational level, this translates into a seamless combination of the upfront and infused strategy and the potential setup of a new project role for a human-centric requirements engineer. The activities comply with both Design Thinking and Requirements Engineering elements to establish an end-to-end view from exploring a user's need to conceptualizing a solution vision and specifying a functional system. When starting a project, the upfront strategy can be used to provide clarity about the problem context and to elicit (user) requirements in a structured yet creative manner. A high-resolution prototype can help to specify the functionalities of the system vision. When moving on to the more technical side of requirements specification, an ad hoc usage of Design Thinking methods can still be initiated in case features are not defined well enough from a user point of view for example.

Roles: The instantiation of a new role incorporates Design Thinking expertise as well as Requirements Engineering expertise and mediating between both schools of thoughts. In this strategy, it is of great importance that the new role can react quickly when choosing methods and artifacts. The role enables the team to work toward a final product in incremental steps. The responsibilities of the project team during this strategy are like the preceding ones as the continuous strategy combines the two other strategies. The team plans and executes the activities to define the final system. The project sponsor has similar responsibilities as described in the previous sections.

Outcome: The continuous strategy results in a comprehensive set of Design Thinking and Requirements Engineering artifacts as shown in Fig. 9. The requirements specification and system design are based on and traceable to customer needs derived from the context specification.

Case Example

Gamma Energy is a large energy provider with subsidiaries worldwide. A diverse project team applied an upfront Design Thinking approach to explore the potential of platforms in the utility sector. The outcome was a solution vision for a digital home improvement platform to advance lead generation. To ensure a human-centered mindset throughout specification and development, a new role was established to use selected Design Thinking tools for enhancing the prototype and filling the backlog with new features. Produced Design Thinking artifacts included high-fidelity prototypes with usability- and feature-oriented test feedback and new solution ideas. Scrum became the guiding framework for development, which enabled the entire project team to work in sprints. During development Design Thinking prototypes were used as boundary objects to enhance communication with relevant internal stakeholders and to foster a human-centered mindset within the team (see Fig. 13).[6]

[6]This case has also been published in Hehn et al. 2020 and Hehn 2020.

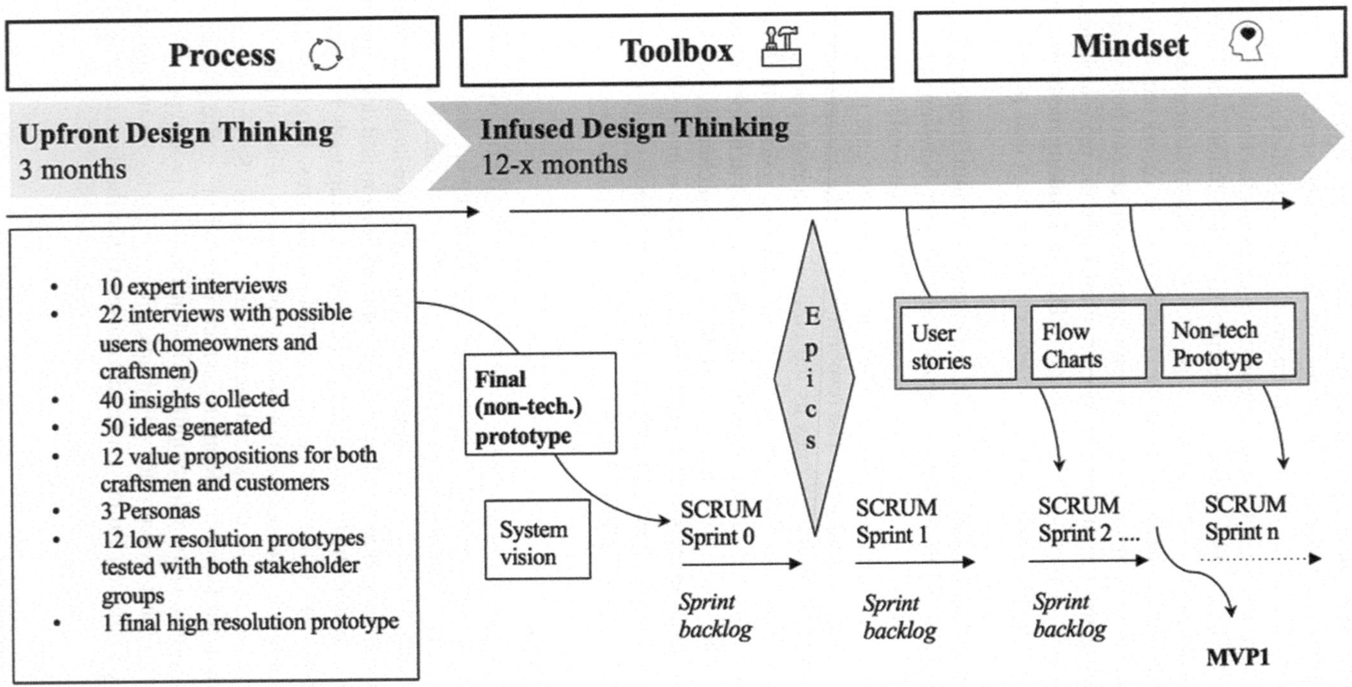

Fig. 13 Evolution from Process, via Toolbox, to Mindset (visualization based on findings from Hehn and Uebernickel 2018, see also Hehn 2020)

Discussion

Our presented operationalization strategies reflect the ongoing discourse of describing Design Thinking at different levels in software engineering approaches (e.g., Brenner et al. 2016; Dobrigkeit and de Paula 2019). In line with other authors, we suggest that the way in which Design Thinking should be used depends on the specific context and objectives of a project. Accordingly, three different strategies with different Design Thinking formats (e.g., process phases, workshops, and single methods) were suggested which are similar to other proposed strategies in research in the context of (agile) software development. Depending on the situation each operationalization strategy offers different benefits but also challenges. Table 4 discusses both for each strategy.

Besides the project context, the existing maturity level of Design Thinking within an organization can be considered an influencing factor when choosing the "right" strategy. While Requirements Engineering is usually an established practice in

Table 4 Benefits and challenges of each operationalization strategy (see also Hehn et al. 2020; Hehn 2020)

Strategy	Benefits	Challenges
Upfront design thinking	– The full potential of Design Thinking is leveraged while changes to Requirements Engineering are not necessary – Due to the focus on problem exploration deep context understanding is achieved – The solution concept has traceable links to user needs	– Resource- and time-intense – Lost (implicit) knowledge and potential starvation of results when handing over Design Thinking results – Little attention is paid to further development critical artifacts such as quality requirements, system constraints, or data models
Infused Design Thinking	– Intervention requires only minimal changes in existing Requirements Engineering practices – Resource and time friendly due to ad hoc usage of selected tools (especially compared to upfront approach) – Low adoption hurdle for Design Thinking methods	– Risk of neglecting problem understanding (especially compared to the upfront approach) – No embedding of Design Thinking mindset due to situational Design Thinking workshops – Little attention is paid to further development critical artifacts such as quality requirements, system constraints, or data models
Continuous Design Thinking	– Seamless integration into existing Requirements Engineering practices including development of critical artifacts – High likelihood of infusing a human-centered mindset within the project team – Precise and traceable (user) requirements through continuous identification of new requirements and testing	– Requires commitment, resources, and time to develop continuous integration of both approaches in an organization – Continuous Design Thinking is highly dependent on the staffing of the project team – Requires an organizational mind shift and support, potentially even an organizational restructuring

industry, Design Thinking is still relatively new. The decision to integrate the two approaches also depends on the level of courage, given time, and dedicated resources. As a rough guideline, the infusion strategy provides a reasonable starting point as it applies focused Design Thinking interventions within established practices. While the upfront strategy also keeps existing procedures, it requires more time and resources. Finally, the continuous strategy demands for a commitment from management to foster mindset change in an organization or department.

A "morphing nature" of Design Thinking in software-intensive development projects can be stipulated, evolving from process guidance, via toolbox support to the manifestation of a human-centered mindset of the project team. When approaching "wicked" problems, Design Thinking starts with a structured, upfront approach to define a clear product vision. Then, it transforms into a loose bundle of tools and a mindset that links well to common agile practices. Figure 13 visualizes this evolution.

Synthesis of Findings

The following sections summarize our findings from sections "An Integrated Artifact Model Combining Design Thinking and Requirements Engineering" and "Operationalization Strategies."

Leveraging the Best of Both Worlds

Design Thinking and Requirements Engineering are not mutually exclusive but rather reinforce and complement each other. Using Design Thinking for Requirements Engineering means putting more focus on the early phases of the process to determine customer needs, requirements, and context, which affects the system vision with its product features and functionalities. Design Thinking expands the toolbox for Requirements Engineering by emphasizing artifacts for defining the relevance of the system vision. It fosters a holistic exploration of the problem context and defines precise user requirements. A prototype shapes the vision of the system. These artifacts complement the more technical-oriented artifacts from Requirements Engineering with a human-centered perspective. In addition, Requirements Engineering expands the toolbox of Design Thinking by connecting Design Thinking artifacts to later-staged software development processes. In this sense, Design Thinking related artifacts are transformed into functionalities for technical realization. What counts in the end in Requirements Engineering is the set of elaborated requirements, while in Design Thinking, not only the prototype is the ultimate outcome, but the learning curve leading to it.

For creating a lasting impact of the system vision on the upcoming design and implementation activities, a balance should be found between the benefits of early experimentation as done in Design Thinking and the advantages of institutionalizing a proper structure and documentation for subsequent software engineering activities as achieved by Requirements Engineering.

A Comprehensive Blueprint for Innovative Software-Intensive Systems

We contribute an evaluated artifact model for Design Thinking and Requirements Engineering that can be tailored to specific project situations. The model is descriptive and prescriptive at the same time. It is descriptive by depicting the most common Design Thinking and Requirements Engineering artifacts as used in (innovative) software-intensive development projects. We see the model as a blueprint for designing new innovative systems, which makes it also prescriptive as it provides a guideline and orientation for generating the artifacts in development projects. Managers can use the model to evaluate their Requirements Engineering processes and, thereby, improve effectiveness and create solutions in a more human-centered fashion.

There is No "One Size Fits All"-Integration Strategy

Operation modes that integrate Design Thinking into (agile) software development approaches have been proposed before (e.g., Lindberg et al. 2012; Häger et al. 2015; Dobrigkeit et al. 2018). Building on these findings and triangulating them with empirical data from industry, three operationalization strategies have been identified in which Requirements Engineering can profit from Design Thinking and vice versa: (1) Run Design Thinking upfront to Requirements Engineering practices (upfront Design Thinking), (2) infuse Requirements Engineering with selected Design Thinking tools (infused Design Thinking), or (3) apply Design Thinking and Requirements Engineering continuously in a flexible manner (continuous Design Thinking). The decision on which strategy to follow depends on the project context and objective. The first strategy is recommended when the problem and solution space is unclear. Here, the Design Thinking process provides a guiding structure for requirements elicitation and the specification of a solution vision. The second strategy offers requirements engineers a way to make use of selected Design Thinking methods when they feel it is necessary. Typically, these are situations in which project members face difficulties in an ongoing Requirements Engineering process that might be addressed by Design Thinking methods. The third strategy supports a continuous yet flexible application of the Design Thinking process and ad hoc tools. The continuous approach entails the evolution from using Design Thinking as a guiding process to applying it as a toolbox for adaptive support up to implementing Design Thinking principles in the mindset of project members. This strategy should be chosen when a sustainable integration of both Design Thinking and Requirements Engineering is intended and the project requires a continuous integration of users into the development project. In this context, the human-centric requirements engineer is a new role that incorporates skills from both disciplines. Business analysts may leverage Design Thinking to deeply explore the system context while design thinkers may equip themselves with Requirements Engineering knowledge to better connect their results to subsequent software design.

Conclusion

Design Thinking offers great potential for promoting innovative, user-centered concepts as it promises to place users and their needs at the core of the design process. This gave rise to great interest in using Design Thinking for the engineering of software-intensive systems and services which are nowadays challenged by their pervasive nature, ever-growing complexity, and the inherent difficulty to capture requirements and development constraints in a user-centric manner. Despite the popularity of Design Thinking in research and practice, it is, however, often treated in isolation without much care for a clear, seamless integration into established software engineering approaches. In fact, too often, we tend to pretend that problem-solving ends with a deeper understanding of the problem domain and by building mostly nontechnical prototypes and, thus, leaving open an effective transition into actual development and quality assurance. At the same time, in software engineering research and practice, we pretend too often that requirements are just there and that they simply need to be elicited and documented (if at all) and, thus, missing out great potential of fully exploring the problem space in a human-centric manner.

The idea of integrating Design Thinking into Software Requirements Engineering approaches to leverage the potential of a deeper problem exploration and discovering and specifying requirements more thoroughly is not new. However, Requirements Engineering and Design Thinking come both in various forms and in interpretations rendering such an integration cumbersome. Thus, integration efforts typically end at the high level of abstract principles, values and mindsets, and practices. In this chapter, we, therefore, took an artifact-centric perspective to (1) synthesize both at a terminological and conceptual level, and to (2) lay the foundation of effectively guiding the problem-oriented specification of requirements based on a seamless and holistic underlying artifact model. Our contributions focus on the following two aspects:

- We elaborated on the very fundamental principles and practices of both Design Thinking and Requirements Engineering and established two independent artifact models that reflect those principles. Here, we drew from both the state of the art in Design Thinking and in Requirements Engineering as well as from experiences made along two decades of academic–industry collaborations.
- We integrated both, the artifact model for Design Thinking and the model for Requirements Engineering and presented different operationalization strategies of how to make efficient use of that integrated approach to create human-centered software-intensive systems.

Note that rather than merely focusing on a purely academically oriented model, we aimed at elaborating on essential terms, principles, and concepts while considering and extending the perspective on the practical relevance as many results emerge from academia–industry collaborations. The choice of the artifact-centric, process-agnostic philosophy further served two major purposes. First, it allowed us to lay such conceptual and terminological foundation for an integrated approach while, second, not enforcing a rigid, pre-defined structure for one (and only one) specific way of working (and thinking), hence, accommodating the various project situations and disciplinary backgrounds we face.

This is in tune with the overall scope of this book. We aim at creating a space to further foster debates and efforts in integrating both Design Thinking and Software and Systems Engineering by inviting scholars and practitioners from both interdisciplinary communities while not enforcing respective historically grown worldviews on each other. One hope we associate with this introductory chapter as well as with the overall book is to motivate the value of such an integration of both worlds.

Acknowledgments We would like to thank Falk Uebernickel for his continuous support and feedback in previous articles and research efforts that provided a major influence on our findings presented in this book chapter. We further thank Manfred Broy and Walter Brenner for stimulating discussions and feedback on earlier versions of this manuscript.

Appendix

Artifact Description

The following appendix defines the content model of the combined artifact model in detail giving for each content item a definition of the used concepts.

The *Number (#)* references the assigned number within the artifact model.

The *Name* captures the name and the type of the artifact. If the artifact can be attributed to both Design Thinking (DT) and Requirements Engineering (RE), different descriptions for both approaches (e.g., Design Challenge and Project

Scope) are marked by a slash (/). In this case, the description for the Design Thinking-related artifact is provided first and the Requirements Engineering expression second.

Description and Purpose denotes the content and main characteristics of each artifact type. Interdependencies summarize the relationships between the artifacts regarding their content within the artifact model. The description differentiates between the input that artifacts receive from the content of other artifacts ('input from') and the output that they provide for other artifacts in the artifact model ('input for').

The *Notation* suggests appropriate documentation and specification techniques for each artifact (e.g., natural language, Unified Modelling Language (UML) class diagrams, model-based documentation).

Context Specification

A description of the content items of the context specification is provided in Table 5.

Requirements Specification

A description of the content items of the requirements specification is provided in Table 6.

System Specification

A description of the content items of the system specification is provided in Table 7.

Table 5 Content items in the context specification

#	Name	Description and purpose	Notation
01	Design Challenge/ Project Scope (DT&RE)	Describes the business problem and provides direction for problem analysis and development; has an exploratory character in DT, a convergent objective in RE Input for (#05), (#08), (#27), (#30)	Natural text
02	Constraints and Rules (DT&RE)	Restrictions and fixed design decisions that influence the system design and implementation and must be obeyed or satisfied; establishing them helps to run and manage the project within the intended business and technical restrictions; constraints are often explicitly challenged in DT Input for (#05)	Natural text
03	Business Case (DT&RE)	Provides rationale for a design project and is used to convince decision maker or project sponsor; in DT its main objective is to evaluate available execution budget (resources and time), in RE it may have concrete solution options in mind Input from (#01); input for (#05)	Natural text
04	Stakeholder map/stakeholder model (DT&RE)	List of relevant stakeholders (internal and external) for the project, typically including project sponsor or client, project manager, product manager, other (senior) decision-makers, investors, end users, customers, operators, product disposers, sales and marketing, or regulatory authorities; helps to identify key internal and external stakeholders as sources of requirements Input from (#01); input for (#05), (#07), (#25)	Natural text, diagram, UML actor hierarchy
05	Objectives and goals (DT&RE)	Prescriptive statements of intent regarding business, usage, or system goals issued by a stakeholder (e.g., quality-related, optimization-specific, behavioral, anti-goals); provide direction for problem analysis and system development; in DT the list contains mainly high-level business goals and objectives provided by the project sponsor to keep outcome and specifics open for exploration; in RE they may be more precise Input from (#01), (#02), (#03), (#04); input for (#06), (#24), (#25)	Natural text, goal graphs

(continued)

Table 5 (continued)

#	Name	Description and purpose	Notation
06	Domain model (RE)	Composed of all real-life conceptual objects related to a specific problem (incl. business entities, attributes, roles, relationships, constraints); ensures an understanding of the landscape of business entities in the problem area and can be used to solve problems related to that domain Input from (03#), (05#); input for (#09, #34, #40), (#24), (#25)	UML activity diagrams; Business Process Model Notation (BPMN)
07	Design space map (DT)	Overview of knowledge and knowledge gaps in the context of the project; helps to structure the exploration phase and provides a common understanding of the design challenge; it evolves over the duration of a project in which new knowledge is added Input from (#01), (#04); input for (#10), (#11)	Natural text
08	Assumptions (DT)	Hypotheses about project and stakeholders to be explored and tested in the project; provides a first overview of possible team biases Input from (#01), (#04); input for (#17), (#18)	Natural text
09	Glossary (RE)	List of all relevant business or technical domain-specific terms to ensure their consistent usage throughout the entire development life cycle; key elements are terms, definitions, aliases, and related terms Input from (#04), (#06), (#07); input for (#34), (#40)	Natural text
10	Secondary research report (DT)	Summary of various sources of information and insights from existing market research about the given subject domain (e.g., market and benchmarking reports, sales reports, internal databases, government statistics, articles, research studies); the report supports the project team to clarify research questions and gain an initial understanding of the challenge context Input from (#07); input for (#12), (#15)	Natural text
11	Field studies (DT)	Collection of raw data (incl. statements, observations, pictures, videos) from interviewees; they help the team to create a common understanding of the raw data and empathize with the interviewees Input from (#04), (#07); input for (#12), (#13), (#14)	Natural text, pictures, videos

(continued)

Table 5 (continued)

#	Name	Description and purpose	Notation
12	Thematic clusters (DT)	Group of user statements, observations, and other findings from primary and secondary research that represent a specific subtopic of the project content; they provide an overview of relevant topics within a given domain and help the project team to recognize patterns Input from (#11), (#10); input for (#15)	Natural text
13	Personas (DT)	Fictional characters that represent a specific stakeholder group relevant to the project (incl. a demographic profile, behavioral patterns, attitudes, goals); they facilitate the understanding of (potential) users' needs, behaviors, motivations, and frustrations and provide alignment for discussing design decisions Input from (#11); input for (#14), (#16), (#17)	Natural text; pictures
14	Customer journeys (DT)	Visual representations of the experience of a customer when interacting with an organization, product, or service (activities, tasks, touchpoints); they offer a systematic analysis of challenges, pain, and gain points that help to identify areas with innovation potential Input from (#11), (#13); input for (#15), (#16), (#25)	Natural text; sequence and activity diagrams
15	Insights (DT)	Findings that occur because of synthesis and interpretation of primary research; usually expressed in one sentence to explain why something is happening Input from (#11), (#12); input for (#16), (#17)	Natural text
16	Opportunity Areas (DT)	Potential for innovation based on insights and needs found in primary research; they define specific directions for next steps while they often go beyond the project assignment itself; the formulation of opportunity areas is rather action-oriented, while the insights describe the status quo or a desired future state Input from (#12)–(#15); input for (#17)	Natural text
17	Solution ideas (DT)	Specific features and concepts on how to solve a given problem statement (based on creativity techniques and brainstorming) Input from (#09); input for (#18), (#20), (#22)	Natural text

(continued)

Table 5 (continued)

#	Name	Description and purpose	Notation
18	Low-fidelity prototypes (DT)	Tangible and testable artifacts that demonstrate the key functionalities of an idea; examples are paper prototypes, role plays, Wizard of Oz; particularly suitable during the early stages of a project, when the topic is still abstract or in the process of forming as costs and effort are extremely low, which allows the project team to explore various ideas at once Input from (#17); input for (#18), (#20)	Different forms, mostly in a paper-based format
19	Scope-oriented test results (DT)	Feedback from users and other relevant stakeholders regarding the basic concept of an idea; it helps the team to gain more empathy for their target group and to decide which ideas to keep, to refine, and to drop input from (#18); input for (#20), (#22)	Natural text
20	Medium-fidelity prototypes (DT)	Non-technical prototype showing key features of the target product or service; while low-fidelity prototypes (#18) are useful to inspire new ideas, medium-fidelity prototypes are mainly used to test and refine existing solution ideas; they usually take more effort to build, yet also provide a much more realistic representation of the envisioned behavior and user interface Input from (#17), (#18), (#19); input for (#21), (#22)	Different forms, mostly in a digital format
21	Feature-oriented Test Results (DT)	Feedback from users and other relevant stakeholders regarding key features and functionalities of the prototype; they validate customer's expectations and help to prioritize functionalities for implementation Input from (#20); input for (#22), (#25, 26)	Natural text

Table 6 Content items in the requirements specification

#	Name	Description and purpose	Notation
22	High-fidelity prototypes (DT)	Offers a clear vision of how the final system will look and feel; they help the project team to gain meaningful feedback for usability testing and are also suitable to gain buy-in from clients and internal project stakeholders Input from (#17), (#18)–(#21); input for (#22), (#24)	Different forms, mostly in a digital format
23	Usability-oriented Test Results (DT)	Feedback from users and other relevant stakeholders regarding the interaction with a product; the results provide areas for improving issues of understandability and point at directions for refining design elements and interaction mechanisms Input from (#22); input for (#24), (#25)	Natural text; pictures, videos
24	System Vision (DT&RE)	Specification of how an information system is to fit into the business context while supporting pre-defined restrictions and goals; it serves as a means for agreeing on what the solution is about; while the purpose of the system vision is similar to both DT and RE, its realization might be different: In DT it is usually comprised of a high-level natural text specification and a medium-or high-fidelity prototype (#20, #22), in RE the system vision is often expressed via rich picture. Input from (#03), (#04), (#05), (#06), (#22); input for (#25), (#33), (#31)	Rich picture, prototype, natural text
25	Usage model (DT&RE)	Illustration of the (black box) system behavior of the system vision (#24) from the user's point of view through an overview of use cases (incl. actor, task, objective, and causal relationship); the model provides an understanding about which system functions are performed for which actors (in their roles); while the purpose of the usage model is similar to both DT and RE, its realization might be different Input from (#13), (#14), (#24), (#26); input for (#28), (#29), (#33)	Natural text, UML activity diagrams
26	Service model (DT&RE)	Specification of requirements and objectives of the intended services of the solution (i.e., user-visible functions through input/output-relations); it provides a comprehensive understanding of the services and their underlying resources and processes, whether seen or unseen by the user; while the purpose of the usage model is like both DT and RE, its realization might be different Input from (#24); input for (#25), (#29), (#33)	Natural text; graphs

(continued)

Table 6 (continued)

#	Name	Description and purpose	Notation
27	Process Requirements (RE)	Activities that should be performed by the developing organization (e.g., compliance to standards and process models, project milestones, style guides, infrastructure); they provide the guidelines for a consistent design and implementation of the intended system Input from (#01)	Natural text
28	Functional hierarchy (RE)	Specification of functions and subfunctions and their relationships and dependencies; functions are user-visible pieces of the system behavior that correspond to services in (#26) and realize system actions from (#25); bridges the requirements and system specification and can be used as a guideline for obtaining and organizing system requirements Input for (#29), (#36), (#38), (#39)	Graphs and input–output tables
29	Data model (RE)	Summary of all data objects and relations that are part of the system's functions and interaction scenarios; it supports the development of the intended system by providing the definition, format, and structure of the required data Input from (#25), (#26), (#28); input for (#37)	UML class diagrams
30	Deployment requirements (RE)	Description of demands for making the software available for use, i.e. specifying the process of the deployment and the technical infrastructure during the initial release of the system or specific parts of it; they contribute to the overall quality of the resulting system Input from (#01)	Natural text
31	Risk list (RE)	Description of all risks that are related to project-specific requirements and that potentially threaten the development or operation of a system; risks are typically analyzed along stakeholder interests and estimated regarding their probability and potential damage; the risk list provides the foundation to introduce necessary countermeasures Input from (#24)	Natural text
32	System constraints (RE)	Logical and technical restrictions for the system architecture, its functionality, and quality; they provide the boundaries for development and deployment Input for (#38)	Natural text
33	Quality requirements (RE)	Desired quality characteristics of a system beyond functionality and features (e.g., reliability, performance, security, usability, adaptability); they are assessed by pre-defined measurements and help to validate the successful completion of an entire system or its respective functions and features Input from (#11), (#13), (05#), (#24), (#25); input for (#36), (#38)	Natural text

(continued)

Table 6 (continued)

#	Name	Description and purpose	Notation
34	Glossary (RE)	Extends the glossary of context-relevant terms (#09) with requirements-specific terms; it will show up again in the system specification (#40) as more terms are added	Natural text
35	Architecture overview (RE)	Aggregation of component overview (#38) and functional hierarchy (#28); offers high-level understanding of the evolving system's architecture and guides the definition of the more intricate functional and operational architecture Input for (#36), (#38)	Component diagram

Table 7 Content items in the system specification

#	Name	Description and purpose	Notation
36	Function model (RE)	Overview diagram of the user-observable functions and their communication relationships; the model ensures an overview of all functions and processes and, thus, assists in determining the scope for implementation and the product and service costs Input from (#28), (#33), (#35), (#38); input for (#39)	Graphs, tables
37	Data model (RE)	Overview of the coarse-grained data objects and the relations that are required for the executing the system's functions; the "data elements" of the data model refine the "data objects" from the data model (#29) in the requirements layer by using a particular data type; it is part of a stepwise completion from moving the focus on defining user-visible functions towards specifying the design system Input from (#29); input for (#39)	UML class diagrams
38	Component model (RE)	Description of the components (i.e., building blocks) of a system's services and their respective channels and interfaces (e.g., application components, system software components, technical components, hardware components); the model bridges the requirements layer with the system layer by defining the main design principles and overall structure of the system Input from (#32), (#33), (#35); input for (#36), (#39)	Component diagrams
39	Behavior model (RE)	Description of the internal behavior of a system with the goal to execute the defined functionalities; the model depicts a dynamic view of the system behavior and illustrates how objects or system components interact to support use cases Input from (#25), (#36), (#38); input for (#37)	Inter-action diagrams, behavioral state machines
40	Glossary (RE)	Extends the previously defined glossary artifacts (#09, #34) with technical relevant terms	Natural text

References

Beyhl T, Giese H (2016) Connecting designing and engineering activities III. In: Plattner H, Meinel C, Leifer L (eds) Design thinking research, understanding innovation. Springer-Verlag, Cham, pp 265–290

Brenner W, Uebernickel F, Abrell T (2016) Design thinking as mindset, process, and toolbox. In: Brenner W, Uebernickel F (eds) Design thinking for innovation: research and practice. Springer, Cham, pp 3–21

Brown T (2008) Design thinking. Harv Bus Rev 86:84–92

Brown T (2009) Change by design, how design thinking transforms organisations and inspires innovation. HarperBusiness, New York

Brown T (2012). Design Thinking defined. https://designthinking.ideo.com/. Accessed 12 Jan 2021

Broy M (2006) Requirements engineering as a key to holistic software quality. In: Proceedings of the 21th international symposium on computer and information sciences. Springer, New York, pp 24–34

Buchanan R (1992) Wicked problems in design thinking. Des Issues 8(2):5–21

Dobrigkeit F, de Paula D (2019) Design thinking in practice: understanding manifestations of design thinking in software engineering. In: Proceedings of the 27th ACM joint European software engineering conference and symposium on the foundations of software engineering, Tallinn, Estonia. ACM, New York, pp 1059–1069

Dobrigkeit F, de Paula D, Uflacker M (2018) InnoDev - a software development methodology integrating design thinking, scrum and lean startup. In: Plattner H, Meinel C, Leifer L (eds) Design thinking – research looking further: design thinking beyond solution-fixation. Springer-Verlag, Cham, pp 199–228

Doorley S, Holcomb S, Klebahn P, Segovia K, Utley J (2018) Design thinking bootleg. d.school at Stanford University, Stanford, CA

Emam KE, Koru AG (2008) A replicated survey of IT software project failures. IEEE Softw 25(5): 84–90

Forrester (2018) "The Total Economic Impact™ Of IBM's design thinking practice. How IBM drives client value and measurable outcomes with its design thinking framework" A Forrester Total Economic Impact™ study, commissioned by IBM

Fraser H (2011) Business design: becoming a bilateral thinker. Rotman Magazine, Winter, pp 70–76

Gutzwiller T (1994) Das CC RIM-Referenzmodell für den Entwurf von betrieblichen, transaktionsorientierten Informationssystemen. Physica, Heidelberg

Häger F, Kowark T, Krüger J, Vetterli C, Uebernickel F, Uflacker M (2015) DT@Scrum: integrating design thinking with software development processes. In: Plattner H, Meinel C, Leifer L (eds) Design thinking research, understanding innovation. Springer-Verlag, Cham, pp 263–289

Harte R, Glynn L, Rodríguez-Molinero A, Baker PM, Scharf T, Quinlan LR, Ólaighin G (2017) A human-centered design methodology to enhance the usability, human factors, and user experience of connected health systems. JMIR Hum Factors 4(1):e8

Hehn J (2020) The use of Design Thinking for a human-centered requirements engineering approach. Dissertation, University of St. Gallen, Baier Druck, Heidelberg

Hehn J, Uebernickel F (2018) The use of Design Thinking for requirements engineering – an ongoing case study in the field of innovative software-intensive systems. In: Proceedings of the 26th IEEE international requirements engineering conference (RE'18), Banff, Canada

Hehn J, Uebernickel F, Stöckli E, Brenner W (2018a) Towards designing human-centered information systems: challenges in specifying requirements in Design Thinking projects. In: Proceedings of the Multikonferenz Wirtschaftsinformatik (MKWI 2018), Lüneburg, Germany

Hehn J, Uebernickel F, Herterich M (2018b) Design Thinking methods for service innovation – a Delphi study. In: Proceedings of the 22nd Pacific Asia conference on information systems (PACIS 2018), Yokohama, Japan

Hehn J, Mendez D, Uebernickel F, Brenner W, Broy M (2020) On integrating Design Thinking for a human-centered requirements engineering. IEEE Software, special issue Design Thinking, pp 25–31

IDEO.org (2015) Field guide to human centered design. http://www.designkit.org/resources/1. Accessed 3 Jan 2019

Inayat I, Salim SS, Marczak S, Daneva M, Shamshirband S (2015) A systematic literature review on agile requirements engineering practices and challenges. Comput Hum Behav 51:915–929

Jönsson P, Lindvall M (2005) Impact analysis. In: Aurum A, Wohlin C (eds) Engineering and managing software requirements. Springer-Verlag, Berlin Heidelberg, pp 117–142

Kolko J (2015) Design thinking comes of age. Harv Bus Rev 93(9):67–71

Kröper M, Lindberg T, Meinel C (2010) Interrelations between motivation, creativity and emotions in Design Thinking processes – an empirical study based on regulatory focus theory. In: Proceedings of the 1st international conference on design creativity, Kobe, pp 97–104

Lauenroth K (2018) Digital design manifesto: a self-confident design profession is the key to successful and sustainable digitalization. Bitkom, Berlin. https://www.digitaldesign.org/content/1-home/digital-design-manifesto.pdf. Accessed 8 Nov 2019

Lindberg T, Köppen E, Rauth I, Meinel C (2012) On the perception, adoption and implementation of Design Thinking in the IT industry. In: Plattner H, Meinel C, Leifer L (eds) Design Thinking research, understanding innovation. Springer-Verlag, Cham, pp 229–240

Maguire M, Bevan N (2002) User requirements analysis. In: Hammond J, Gross T, Wesson J (eds) Usability. Springer, Boston, MA, pp 133–148

Martin R (2009) The design of business. Why Design Thinking is the next competitive advantage. Harvard Business Review Press, Boston, MA

ME 310 (2010) ME310 design innovation at Stanford University. Micro Cycle. https://web.stanford.edu/group/me310/me310_2016/. Accessed 13 Jan 2019

Mendez Fernandez D, Penzenstadler B (2014) Artefact-based requirements engineering: the AMDiRE approach. Requir Eng J 20(4):405–434

Mendez Fernandez D, Wagner S (2014) Naming the pain in requirements engineering: a design for a global family of surveys and first results from Germany. Inf Softw Technol 57:616–643

Mendez Fernandez D, Wagner S, Lochmann K, Baumann A, de Carne H (2012) Field study on requirements engineering: investigation of artefacts, project parameters, and execution strategies. Inf Softw Technol 54(2):62–178

Mendez Fernandez D, Wagner S, Kalinowski M, Schekelmann, Tuzcu A, Conte T, Spinola R, Prikladnicki R (2015) Naming the pain in requirements engineering: comparing practices in Brazil and Germany. IEEE Softw Voice Evid 32(5):16–23

Mendez Fernandez D, Wagner S, Kalinowski M, Felderer M, Mafra P, Vetrò A, Conte T, Christiansson MT, Greer D, Lassenius C, Männistö T, Nayebi M, Oivo M, Penzenstadler B, Pfahl D, Prikladnicki R, Ruhe G, Schekelmann A, Sen S, Spinola R, de la Vara JL, Tuzcu A, Wieringa R (2016) Naming the pain in requirements engineering: contemporary problems, causes, and effects in practice. Empir Softw Eng J. https://doi.org/10.1007/s10664-016-9451-7

Mendez Fernandez D, Böhm W, Vogelsang A, Mund J, Broy M, Kuhrmann M, Weyer T (2019) Artefacts in software engineering: a fundamental positioning. Softw Syst Model 18:2777–2786

Newman P, Ferrario MA, Simm W, Forshawz S, Friday A, Whittle J (2015) The role of Design Thinking and physical prototyping in social software engineering. In: Proceedings of the 37th international conference on software engineering, Florence, Italy. IEEE, pp 487–496

Przybilla L, Schreieck M, Klinker K, Pflügler C, Wiesche M, Krcmar H (2018) Combining Design Thinking and agile development to master highly innovative IT-projects. In: Mikuzs M, Volland A, Engstler M, Hanser E, Linssen O (eds) Projektmanagement und Vorgehensmodelle 2018 – Der Einfluss der Digitalisierung auf Projektmanagementmethoden und Entwicklungsprozesse. Gesellschaft für Informatik, Bonn, pp 113–124

Robertson S, Robertson J (2013) Mastering the requirements process: getting requirements right. Pearson Education

Schmiedgen J, Rhinow H, Köppen E, Meinel C (2015) Parts without a whole? – The current state of Design Thinking practice in organisations. Technische Berichte des Hasso-Plattner-Instituts für Softwaresystemtechnik an der Universität Potsdam, Study report no. 97

Schön DA (1984) The reflective practitioner: how professionals think in action. Basic Books, New York

Soledade MP, Freitas R, Peres SM, Fantinato M, Steinbeck R, Araújo U (2013) Experimenting with design thinking in requirements refinement for a learning management system. In: Anais do Simpósio Brasileiro de Sistemas de Informação, pp 1–13

Uebernickel F, Brenner W, Naef T, Pukall B, Schindlholzer B (2015) Design thinking: das Handbuch. Frankfurter Allgemeine Buch, Frankfurt

Venkatesh Sharma K, Kumar PV (2013) A method to risk analysis in requirement engineering using tropos goal model with optimized candidate solutions. Int J Comput Sci Issues 10(6):250–259

Vetterli C, Brenner W, Uebernickel F, Petrie C (2013) From palaces to yurts: why requirements engineering needs design thinking. IEEE Internet Comput 17(2):91–94

Wagner S, Méndez Fernández D, Kalinowski M, Felderer M, Mafra P, Vetrò A, Conte T, Christiansson MT, Greer D, Lassenius C, Männistö T, Nayebi M, Oivo M, Penzenstadler B, Pfahl D, Prikladnicki R, Ruhe G, Schekelmann A, Sen S, Spinola R, de la Vara JL, Tuzcu A, Wieringa R, Winkler D (2019) Status Quo in requirements engineering: a theory and a global family of surveys. ACM Trans Softw Eng Methodol 28(2):9

Yoo Y (2017) Design thinking for IS research. MIS Q 4(1):iii–xviii

From Design Thinking in Software Engineering to Digital Design as a New Profession: An Essay on Methods and Professions for Shaping Digital Solutions and Systems

Kim Lauenroth

Experiences with Design Thinking in Software Projects

My personal journey to the country of Design Thinking started more than 10 years ago with the documentary "Objectified" from Gary Hustwit.[1] This documentary is dedicated to the profession of industrial design. Famous designer, e.g., Jonathan Ive from Apple or Dieter Rams from Braun give deep insights into their daily work and show how designers think about ideas and shape products.

What puzzled me with this documentary was, that software was only a minor part of this movie. Bill Moggridge, one of the founders of the Design Thinking company IDEO, briefly talked about software (Hustwit 2015). But not in the same way as the other designers talked about their products. However, I did not give much attention to this observation at that time. The documentary was a pointer to Design Thinking as a keyword and was an important motivation to learn more about this topic.

There are many understandings of Design Thinking out there in practice. Some people call it a mindset, other call it a method or a framework. My understanding of Design Thinking has been shaped by Tim Brown's book (Brown 2009). I understand Design Thinking as a framework based on the human-centered design process that provides a rich body of techniques that can be applied within this process.

Now, 10 years later, I have seen a broad spectrum of successes and failures which can be summarized as follows:

[1] https://www.hustwit.com/objectified

K. Lauenroth (✉)
adesso SE, Dortmund, Germany

University of Applied Science and Arts Dortmund, Dortmund, Germany
e-mail: kim.lauenroth@fh-dortmund.de

J. Hehn et al. (eds.), *Design Thinking for Software Engineering*, Progress in IS,
https://doi.org/10.1007/978-3-030-90594-1_3

- *Design Thinking for scoping and framing an innovative software product works very well.* When starting a project for an innovative software product, Design Thinking has supported me very well in creating an environment in which stakeholders and a first software team can develop a mutual understanding of the scope and the overall vision of the core innovation ideas of the project. The results of such Design Thinking workshops were a great input for a team of requirements engineers who structured the results and then developed a first specification of the product together with the relevant stakeholders.
- *Small, clearly focused Design Thinking workshops are a great tool to tackle "small" problems.* In my professional live, I am often responsible for the requirements part of larger software projects. This includes defining the process for working on requirements and leading a team of requirements engineers. In some situations, our clients feel the need to find innovative solutions to certain parts of software systems. In such situations, Design Thinking as a timeboxed workshop with relevant stakeholders (especially potential end-users) and a clearly defined scope was always a very efficient way of finding appropriate solution ideas.
- *"We need innovation in our software, let's do Design Thinking" does not really work.* I have also seen a lot of organizations that maintain software systems in larger companies. Sometimes, such organizations feel an abstract need for innovation in their software. Design Thinking seemed to be a good approach for this situation. Design Thinking workshops came up with some promising improvement ideas for the software and the prototypes created in workshops got positive feedback from all relevant stakeholders. Now two things happened most of the time. Firstly, several promising ideas were discarded at once in the further development process because of feasibility or budget reasons. Secondly, ideas that were finally implemented did not meet the expectations that were created during the Design Thinking workshop.

From an empirical researcher's perspective, my negative experiences should certainly be viewed critically. Many causes may have led to the fact that the ideas could not be successfully implemented. Nor should the fault be seen exclusively in Design Thinking.

Nevertheless, I have met many people in my professional practice who have had similar experiences. The invitation letter from the editors of this book somehow seems to reflect my experiences as well: Integrating Design Thinking more deeply into software engineering is somehow difficult. But the question is why? In this essay, I will give my personal explanation for this situation and draw some conclusions for future activities in the area of Design Thinking in software engineering.

Design Thinking is Rooted in Industrial Design

Let us come back to my observation from the documentary Objectified. I assume, the industrial designers from the documentary did not really talk about software because they do not consider it a relevant part of their job. They design physical products and not software. And here the starting point for the issues that I see in the application of Design Thinking with respect to software.

According to Brown (2009), Design Thinking has been derived from the working style of industrial designers and product designers.[2] There is one implicit assumption that is deeply embedded into these design professions, design is about defining a product's form and function in advance of the physical act of making (realizing) it in (often automated) production processes (de Noblet 1996). The double diamond design process model always ends with testing prototypes of solution ideas. Prototyping is even considered a core element of Design Thinking with the Design Thinking mantra "fail fast, fail early" (Brown 2009).

When looking at what the prototype in industrial design means, it is in the end often a kind of pre-production prototype: a final version of the product that is ready for replication. This means, that industrial designers create a perfect original of a product that is then literately copied in a manufacturing mass production process. Speaking in terms of software engineering, such an approach could be categorized as a genuine waterfall process. Every detail is specified upfront and implementation is understood as a pure mechanical production process. We know from industrial practice that this approach was considered impractical from the very early days of software development (Royce 1970) and that the development of software (i.e., writing code) is a creative act in itself (Glass 2006).

From my experience, the industrial design category of prototypes is a core success factor for Design Thinking: creating a prototype which experience is as close as possible to the solution idea to get realistic feedback (Brown 2009). The emphasis here is on experience, not on appearance. This is the reason, why Design Thinking books make excessive use of paper prototypes and other simple material such as cardboard, wire, etc.

I believe, Design Thinking in software projects can become remarkably successful if the scope of the software to be designed (or the underlying problem) is rather small and can be captured in the essence by an industrial design prototype. With small, I do not mean simple or even trivial. The examples given in Brown (2009) show that Design Thinking has the potential to create innovative product ideas. However, compared to the typical scope of software-intensive systems, which are my daily business and are within the focus of this book, the scope of such products is rather small and simple.

[2] In order to improve the readability of this contribution, I use the term "industrial design" as synonym for all product-related design disciplines in the remainder of this contribution.

Experiences on the Limits of Design Thinking for Software Projects: Two Examples

Design Thinking as a Tool Within a Large Software Project

I have been the lead requirements engineer for a software system that was intended to manage and support all business processes within this organization. The system handles the management of over 40 diverse types of business processes with more than 2,500 process executions per day in four diverse types of business units. Overall, the system will be used by more than 10,000 users per day. The system further includes the management and assignment of employees to orders and the planning of orders with optimization of travel times. As a further challenge some use cases of the system must be offline capable, i.e., the function must be available without a connection to the system servers.

Now assume that you take my role at the point in time where this project was planned. There was a small project team, and I had the responsibility to setup the project that captures the essential requirements and creates an initial specification for this system to plan for a development project. Is a Design Thinking approach appropriate in such a situation?

To be honest, we never considered it an appropriate approach for starting this project because the scope was by far too broad to capture it within the framework and tools of Design Thinking. Even the scope of one business unit was by far too large because even on this level, there was a big variance between the different types of business processes. One theoretical option would have been to approach each business process with a Design Thinking team to shape an innovative software solution for this particular business process. However, this option was by no means realistic because of resources.

Now, one might argue that this is a killer example and that such examples are not the norm in software business. I do not know if this project is the norm, but in my experience, most of the software projects I have seen, have a level of complexity that is similar to the project sketched above. Many software systems have a rather broad scope of functionalities. Therefore, I am convinced that Design Thinking as a *general approach* is not applicable for software-intensive systems.

However, we did apply Design Thinking within this project as an approach to tackle parts of the business processes that called for an innovative solution. As mentioned above, some business processes include travel time. This means, a manager had to plan a team of employees to work at a certain location as part of the business process. In the legacy system, there were several solutions to this problem including self-developed software, office software solutions, and even paper-based planning. One goal of our project was to provide an innovative software solution for this planning process. Within a short period of time, a group of stakeholders and project members developed several ideas that eventually evolved to the final solution. From my understanding, Design Thinking worked in this context because the scope of the travel planning problem was small enough to fit within the range that works with Design Thinking.

Unfinished Design Thinking as a Starting Point

In another situation, I was in charge of finding innovative ideas for improving software-based processes in dental practices. At the start of the project, two Design Thinking workshops were held, one with dentists and one with dental assistants. During both workshops, the scope of the prototypes was overly broad. Some prototypes described small improvements of the process. However, most of the prototype idea described completely new software systems that questioned the structure and functionality of the existing software systems for dental practice.

In this situation, we decided to stop the Design Thinking process in both workshops and started to work on the limits of existing software systems to better understand the problems that the dentists and assistants had. From a Design Thinking perspective, this decision may be considered wrong. In hindsight, this approach was the right one for the project.

Afterwards, we were able to develop a completely new process concept for dental software based on the input from both workshops, which solved many of the problems described in the Design Thinking process. I am convinced that continuing the Design Thinking process with further work and discussion on prototypes would have never been able to create such a comprehensive concept. Nevertheless, without the Design Thinking process, it would have been very likely, that we (our client and our team) would have never gained the confidence to considered going into this direction. Therefore, the unfinished Design Thinking provided important insights that were necessary to start exploring the idea of developing a new software from scratch.

Intermediate Conclusion for Design Thinking Research and Practice for Software-Intensive Systems

Despite some negative experiences, I am convinced that Design Thinking is an important and useful tool that is applicable under certain constraints for software-intensive systems. I see two success factors:

1. Size of the problem scope—the problem scope must be small enough to be grasped and processed by the Design Thinking framework
2. Size of the solution idea—the size of the solution idea must be small enough to be captured by the prototyping understanding of Design Thinking

My recommendation for industrial practitioners is to consider both factors with care when thinking about the application of Design Thinking within one's own organization. The first factor can be considered from the very beginning whereas the second factor should be observed during the Design Thinking process. When people start to think about prototypes, the size of the solution ideas should be considered carefully. When the prototypes appear to be in the scope of industrial design prototypes, the results can be promising. When the size of the prototypes appears to be too large, one should even consider stopping the Design Thinking process or decide to reframe the problem to focus on smaller parts of the problem that can be addressed with smaller prototypes. To be honest, I have no clear guidelines for both factors. I rely on my gut feeling and the feeling of my colleagues when we decide for or against Design Thinking in each situation.

From my personal software research experience, I would consider broad studies of both factors to be exciting. First, there is the question of whether the two factors are really as crucial as they seem. On the other hand, it would be interesting to substantiate both factors with empirically proven data.

Who is the Industrial Designer in Software Engineering?

Although everything discussed so far is interesting, it did not feel right to me for quite a while. What is wrong with software engineering or Design Thinking that they only fit to some extend with each other?

Design Thinking is the accumulation of best practices in design. Above, we already discussed that industrial design has a special worldview in the sense that design produces prototypes of products than can be copied by means of mass production. And this paradigm does not really fit with the ways, software is developed. So, this means that it is not a problem of Design Thinking, but a problem of the sources of which Design Thinking has been derived from. Maybe, the solution is to start the process from scratch and go to the people who design software nowadays to generalize their best practices into a new approach. So, there is a very simple question: Who is the industrial designers equivalent for software engineering?

The answer, or at least, my answer to this question can be a surprise: they do not exist! Industrial design is a dedicated profession with a more than 100 years of history, dedicated bachelor and master programs, and professional associations all over the world. According to the SWEBOK (Bourque and Fairley 2014), we have a knowledge area called software design. But there is no real profession with the self-understanding and education represented by industrial design.

Why is There No Profession Like Industrial Design for Software?

This insight was quite a surprise for me because it is so obvious. The design of software should be at least as challenging as the design of physical products. Why then does such a profession not exist? My answer to this question can be surprising as well: We did not need such a profession so far!

My explanation for this is that software has played a subordinate role for most of its existence and that software has only ever had to implement existing ideas. When looking at the history of computer science, history starts in the 1950th. Software Engineering can be dated back to the 1968 NATO conference in Garmisch (Naur and Randell 1969). At that time, software engineering was about controlling machines (e.g., the eagle moon lander) or about processing data that has been previously processed on paper (e.g., bank accounts or insurance policies). This subordinate role has made its way deeply into the methodological structure of software engineering.

Requirements work in terms of requirements engineering and usability engineering is always considered the starting point of every software engineering endeavor (Bourque and Fairley 2014) which for me means that there is the implicit assumption that somebody else outside the software engineering process (the stakeholder) has understood the often called "real-world problem" that the software must solve. It is important to recognize that this gap is not a weakness! It is one of the core strengths of software engineering that makes it an engineering discipline. Engineering good software is a challenge in itself and requires substantial skills (Glass 2006).

But what does this mean for a stronger integration of software engineering and Design Thinking? I believe that the methodological level will only create small improvements like the one indicated in the first part of this contribution. These improvements are valuable, but a substantial improvement for software-intensive systems requires another level of consideration.

We Need a Dedicated Design Profession that is Able to Design with Software!

A brief look at other disciplines is a good starting point for discussion about professions. In the building industry a design-oriented profession (architecture) and an engineering-oriented profession (civil engineering) have appeared because the complexity of buildings requires specialist of both directions. I am convinced that the same is necessary for software.

A general argument against this analogy is that software is different from bricks and concrete. I disagree with this argument. Creating large buildings is as challenging as software engineering and I am convinced that the creation of buildings has more in common with software than other industries. However, this is a topic for another paper. Even if software is more challenging and different from buildings, it is even a stronger argument that we need dedicated design and engineering professions for software.

Why is a profession more important than methods and techniques? Studying methods and techniques is a core part of software engineering research. This is an important work for understanding software engineering. However, methods and techniques are always performed by people. Therefore, people and their capabilities are at least as important as methods and techniques (Lauenroth and Kamsties 2016).

And there is an even more important argument with respect to the discussion on Design Thinking and software: Professions attract people and there is a real difference between people who are attracted by design professions compared to engineering professions. A study presented in Durling et al. (1996) gives some interesting insights into the personality and learning preferences of industrial design students. When looking at this contribution, one could argue that the university education in software engineering does not really attract people who are interested in a design-oriented way of working because they want to enter an engineering profession. Design as understood, for example, in industrial design is considered a different way of working and knowing (Cross 2006). This difference is not a minor one and should not be underestimated. This difference could explain the issues that Design Thinking has in the software engineering world: The design working style does not really match with the personal preferences of the people that are nowadays involved in software engineering.

With this argumentation, I am convinced that the next level for developing software-intensive systems does not only require methodological research in the area of Design Thinking, but also an initiative to establish a dedicated software-oriented design profession that works side by side with software engineering professionals in the ways as architects and civil engineers nowadays work together in the building industry.

Digital Design as a New Profession for Shaping Digital Solutions

An initiative with such a goal was established in 2018 by the German digital association Bitkom: the digital design manifesto (Bitkom 2018). The core idea behind this initiative is to establish a design profession that has the same self-understanding about software and digitalization as architects have about buildings and industrial designers have about products.

From a terminological perspective, it is important to explain the shift from software to the term digital design. The term software design is already defined and is related to the structure of software (Bourque and Fairley 2014). When talking about software-intensive products, services, or systems, the emphasis is put on software. This perspective is important, but not sufficient.

Outside the software world, the terms digital transformation or digitalization are used to refer a phenomenon that is larger than software: transforming economy and society by means of digital technology (e.g., World Economic Forum 2020). Software is an important part of digital technology, but only one part beside business processes, networks, hardware, end user devices, and other technologies. With this in mind, digital design aims at designing digital products, services, solutions, or services in the same way as industrial designers shape physical products or architects shape buildings.

Competence Profile of Digital Designers at a Glance

According to Bitkom (2018), a digital designer has a competence profile with two focus points: Competence in digital material combined with design competence.

Competence in digital material means an understanding of the capabilities and limits of digital technology that is sufficient to shape a digital solution in the same way as other materials (e.g., wood) can be shaped to create physical products (e.g., a chair). This competence in digital material does not necessarily mean software engineering skills. A very simple example can be used to illustrate this.

Assume, you own a pizza restaurant and want to offer a very simple digital way for selling pizza. You could accept and confirm pizza orders with a public instant messaging service. Payment for the order can be made through an online payment service. As soon as the pizza is paid, it can be made. When the delivery driver has left, you then send a short message to your customer to indicate that the pizza is on the way. This is for sure an oversimplified example, but it shows a way of using digital technology that is independent from software development. It also shows that digital technology can be used to shape simple digital solutions (here an online pizza service).

The other focus point for digital design is design competence. Design competence includes a deep understanding of design as a way of working (Cross 2006) including the ability to communicate design ideas by means of concepts and prototypes, design process competence (e.g., understanding of the human-centered design process, and Design Thinking), and an understanding of the integration of digital design into the development process of digital solutions.

Conclusion: Process Competence as a Core Success Factor for Design of Products with Software Aspects

This contribution started with a discussion of firsthand experiences with Design Thinking in software development and continued with the discussion of a new profession called digital design for shaping digital solutions that also include software. In this section, this contribution is concluded with a deeper look into an important aspect of digital design, namely competence in the integration of digital design into the development process.

An understanding of how a product is made is considered an important part in industrial design education (de Noblet 1996). Designers need to understand the capabilities and limits of the material and the production process in order to design products that can make use of the material and that can be produced efficiently and effectively. One important origin of this combination was the Staatliches Bauhaus in Weimar (Gropius 1923) where Walter Gropius developed a new education program for architects and designers that combined design competence with technical competence that was summarized with the motto "art and technology, a new unity." With this combination, the Bauhaus contributed significantly to the foundations for industrial design education (de Noblet 1996).

When looking at software, the development process is very different from the mass production process of physical products. The success of agile development shows two things with respect to software development (Meyer 2014):

1. A software development can start with a rough understanding of the software to developed.
2. Many decisions about the form and function of a software can be delayed until shortly before their implementation.

There are, of course, decisions about software that should be made very early because changing those decisions later becomes very costly. Good examples for such decisions are architectural decisions that drive the performance and scalability of a software system (Meyer 2014).

With this in mind, I see an important weakness in Design Thinking. Design Thinking neglects the special properties of software development processes in terms of level of detail and the timing of design decisions. This does not become a problem as long as Design Thinking is considered separated from software development processes and provides input in terms of ideas, prototypes, etc.

Furthermore, it is from my experience also not sufficient to invite software people into a Design Thinking team. A software engineer in a Design Thinking team can provide interesting impulses for ideas and prototypes. But they cannot anticipate all the detailed questions and thoughts that come up when ideas are actually implemented. If this would be possible, the original waterfall idea would have been successful.

For a deeper integration of Design Thinking in software development, I am not convinced that new process models (e.g., combining Scrum and human-centered design) are a proper answer or a good research direction. From my personal experience, such models can only give a rough guideline.

More important is concrete experience with the parallel design and development of software. In my project work, I made very good experience with people that have practical software development experience and now work as requirements engineers or UX designers. Their hands-on experience in software development gives them a very good gut feeling on questions about the feasibility of ideas, the level of detail required for decisions, and the timing of decisions.

Outlook: Studying Concrete Examples is More Promising than Studying Methods

Architecture and industrial design are full-flavored bachelor/master programs at universities. These programs heavily rely on practical work. Furthermore, graduates need a lot of practical training to become of good architect or industrial designer.

We have to teach people so that they become great designers and engineers for software-intensive systems. They need to be able to decide, depending on the specific situation, how to continue the development process and what the next methodological steps are.

Process frameworks such as Design Thinking are a good starting point to acquire these skills and methodological research is important for understanding such processes. However, I am convinced that learning design (and engineering) practice is not a matter of methodological research.

With the industrial design education in mind, I think that empirical research on concrete example projects and products is at least of equal important. Concrete example projects and products can provide blueprints and be used as a reflection point to study and discuss the work of people. Thereby, students and researchers can learn from the experience of other people in the same ways as architecture or design research and education benefits from the experience of others.

Design is not an end in itself, it is about results. The literature on industrial design is full of examples of great products (de Noblet 1996). In comparison, the literature on software has practically no examples. We should take industrial design and architecture as a model and look more closely at concrete examples. Only in this way can we achieve a new level of understanding.

References

Bitkom (2018) Digital design manifesto. Bitkom. https://www.digital-design-manifest.de/en/. Accessed 10 Aug 2021

Bourque P, Fairley RE (2014) SWEBOK Guide V3.0: guide to the software engineering body of knowledge. IEEE

Brown T (2009) Change by design: how design thinking transforms organizations and inspires innovation. Harper Business, New York

Cross N (2006) Designerly ways of knowing. Springer, London

de Noblet J (1996) Industrial design – reflections of a century. Flammarion, Paris

Durling D, Cross N, Johnson J (1996) Personality and learning preferences of students in design and design-related disciplines. IDATER conference, Loughborough, England

Glass RL (2006) Software creativity 2.0. developer.* Books

Gropius W (1923) Idee und Aufbau des staatlichen Bauhauses Weimar. Bauhausverlag, München
Hustwit G (2015) Helvetica, objectified, urbanized: the complete interviews. Versions Publishing
Lauenroth K, Kamsties E (2016) People's capabilities are a blind spot in RE research and practice. In: Proceedings of the 22nd international working conference on requirements engineering foundation for software quality, vol 9619. ACM, pp 243–248
Meyer B (2014) Agile – the good, the hype, and the ugly. Springer, Cham
Naur P, Randell B (1969) Software engineering: report of a conference sponsored by the NATO Science Committee. https://www.scrummanager.net/files/nato1968e.pdf
Royce W (1970) Managing the development of large software systems: concepts and techniques. In: Proc. IEEE WESTCON, Los Angeles
World Economic Forum (2020) Digital transformation: powering the great reset. http://www3.weforum.org/docs/WEF_Digital_Transformation_Powering_the_Great_Reset_2020.pdf. Accessed 29 Dec 2020

The Hybrid Model: Combination of Big Data Analytics and Design Thinking

Michael Lewrick

The Combination of Big Data Analytics and Design Thinking

It sounds more than logical to combine analytical approaches and creative thinking. However, in the reality of today's business we often experience the opposite. This was the trigger to create a combined process model in which solutions are developed with elements from Design Thinking and Big Data Analytics. After more than a decade of application, our experiences strengthen our confidence that the results and solutions have in many cases produced better solutions. Thus, the hybrid model is not only used in the innovation of products, services, and ecosystems, but has also found widespread acceptance in the training of data scientists. The reason for this is easily found, because Data reveals the "what" and Design Thinking explores the "how" (see Fig. 1). Both are relevant for creating impactful innovations and solving customer/user problems. In 2010, when we started our first experiments with the hybrid model, a data-driven innovation project was one of the test environments to find out how a combination of data analytics and Design Thinking can lead to better results. Later we applied the hybrid model also to Design Thinking projects of all kinds or projects aiming to kick-start digital transformation efforts. The list below provides a quick reference of where, what, when, and for whom the hybrid model might find application:

Where?

Creator of the Hybrid Model and lead-author of the international bestsellers "The Design Thinking Playbook"; "The Design Thinking Toolbox," and "Design Thinking for Business Growth"

M. Lewrick (✉)
Lewrick & Company, Zürich, Switzerland
e-mail: michael.lewrick@hslu.ch

J. Hehn et al. (eds.), *Design Thinking for Software Engineering*, Progress in IS,
https://doi.org/10.1007/978-3-030-90594-1_4

Fig. 1 Data scientist, source Lewrick et al. (2018, p. 312)

- Digitization projects
- Data Science projects
- Design Thinking challenges

What?

- Developing new products services and processes
- Generally solving problems
- Business Ecosystem Design

When?

- When a data basis exists
- When data can be generated in real-time
- When high agility is required

For whom?

- Interface between customer requirements and customer data
- For the whole organization, customers, or other stakeholders

While applying the hybrid model it was soon recognized that the teams in the organization were becoming even more interdisciplinary than in Design Thinking and that the mixed teams were using a different technical language, which initially did not simplify communication. To the well-known group of team members in Design Thinking projects (design leads, implementation managers, project sponsors, business departments heads, product managers, and other management members) a new species of data experts with various specialties becomes part of the wider team. The new team members had job titles like business intelligence analysts, database administrators, data engineers, and data scientists. The biggest challenge has been having the Big Data specialists interacting with team members already trained and purely focused on the Design Thinking mindset. As a result, the hybrid teams become part of the hybrid model which includes also the hybrid mindset, process, and toolbox explained in this contribution. At the same time, it became also apparent that workshop facilitators applying the hybrid model needed new skills to orchestrate the differences and to understand the tools and methods of both, Design Thinking and Big Data Analytics.

Categories of Combined Models

In general, the combination of Big Data and Design Thinking can be divided into three categories, each with different characteristics (see Table 1). The process model described in this contribution refers to "The hybrid model" which was originally developed in a joint research project from academia and large enterprises on the forefront of Big Data Analytics and Design Thinking. Later the model has also found its way in the context of future tools, methods, and mindsets in the "The Design Thinking Playbook" authored by Lewrick et al. (2018). Today the hybrid mindset finds application in many mixed design teams of enterprises, business ecosystem initiatives, and management and IT consulting companies globally.

The initial motivation to combine Big Data and Design Thinking in a hybrid model was based primarily on three hypotheses. Today we know that the combination brings many more advantages and that with AI-enhanced data processing tools many insights can be generated even faster and are partly already available in real time.

- Increase the efficiency and effectiveness of the innovation process
- Combine the deep customer insights (Design Thinking) and the deep learning from data (Data Sciences)
- Generating synergies by combining the rather qualitative aspects of Design Thinking and the quantitative methods of Data Science

Table 1 Categorization of models

Human-centered Data Science	Design Thinking methodology, method, mindset for Data Science, organizations, and projects
Data-enhanced Design Thinking	Gain in efficiency and effectiveness in the Design Thinking process through data analytics incorporated in each step of the Design Thinking process
The hybrid model	Combination of Design Thinking and Data Science process with a mutual enrichment of tools and methods

The Hybrid Mindset

The biggest challenge in the collaboration of design thinkers and data scientists is to establish a suitable mindset on which the team members from the different areas can build. On the one hand, the core idea of Design Thinking should not be lost in such a mindset and, on the other hand, completely new aspects have to be considered, which are relevant for the work of Data Scientists. The mindset elements shown in Table 2 illustrate the breadth and depth that a hybrid mindset requires from mixed teams and at the same time from workshop facilitators guiding those teams from the problem identification to first ideas, prototypes to the final solution. While the Design Thinking mindset is focusing on the collaboration in diverse teams, envisioning a radical new future, embracing uncertainty, and being human-centered—the Data Analytics mindset are driven by iterating with explainable models, modeling utilities and inputs, optimizing for speed of learning and being data-driven.

The best results are known to be achieved when different T-shaped participants work together in hybrid teams, thus realizing the full potential of the hybrid model. The challenge, however, lies in taking the other perspective, i.e. that data scientists learn to take the perspective of design thinkers and vice versa or at least understood that the other approach and the associated procedure brings advantages over the entire cycle from understanding the problem space to finding solutions. Thus, a shared mindset is crucial for collaboration because the respective fundamental beliefs, attitudes, and biases influence how information is processed and the context in which the world is set. However, it seems to be important that the positive and optimistic mindset of Design Thinking is maintained, because a winning perspective and search for market opportunities leads to long-term success.

Table 2 Mindsets: data science, hybrid, and Design Thinking

Mindsets		
Data Science	Hybrid	Design Thinking
Curious		
Innovative		
Sophisticated analytics and creative problem solving		
Experimenting and testing		
Visualizing and storytelling		
Collaborating		
Correlations and indicators	Accepting indicators and ambiguity	Accepting ambiguity
Analytical	Combining analytics and intuition	Intuitive + (analytical)
Quantitative approach	Combining quantitative and qualitative insights	Qualitative approach
Data-driven	Combining data-driven and human-centered reasoning	Human-centered

The Hybrid Teams

In addition to the mindset that provides the overarching frame, collaboration requires appropriate team members who work together in the hybrid teams. In the context of Design Thinking, great importance has always been attached to working in heterogeneous teams (see Fig. 2). The teams often consist of a combination of expertise from business, marketing, sales, strategy, and IT experts. In hybrid teams, the respective data scientists and specialists are added. These include nowadays data engineers, big data engineers, machine learning scientists, business analytics specialists, data visualization developers, business intelligence (BI) engineers, BI solutions architects, BI specialists, analytics managers, machine learning engineers and of course statisticians. Most of the necessary experts are fully or partially available in large organizations, so that hybrid teams can be assembled without any problems. The bigger challenge is to find suitable workshop facilitators for the Hybrid Mindset. Only few workshop facilitators have a comprehensive insight into the methods and tools of both approaches and can use them to guide the teams.

The Hybrid Process

As in Design Thinking, the hybrid process serves only to orient the teams and provides information about where they are in the process. As a basis for the hybrid model, the Design Thinking process of the d.school in functioning as reference, with the phases: Understand, Observe, Point of View, Ideate, Prototype, and Test. For mapping with the common models from the Data Sciences, the Realize phase was

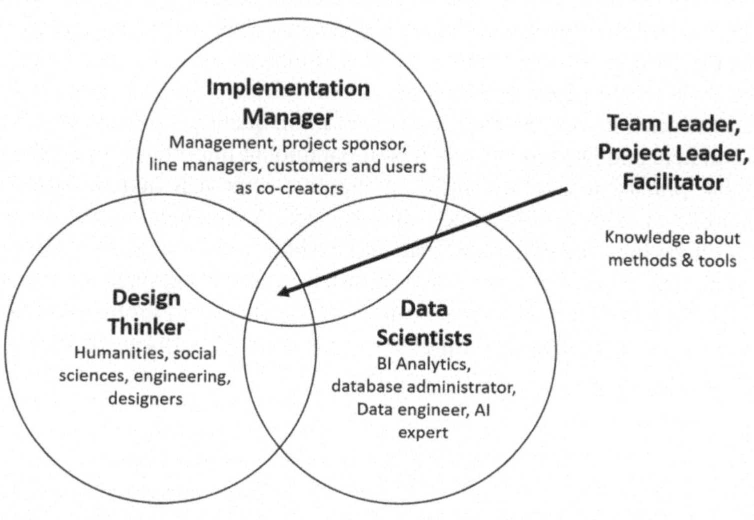

Fig. 2 Composition of a hybrid team

Table 3 Phases in Design Thinking, data science, and in the hybrid model

Model	Phases					
Design Thinking	Understand	Observe	Point of view	Ideate	Prototype	Test	Realize
Data Science	Data content	Data context	Data diversity	Analytics simulate	Analytics agility		Analytics improve
Hybrid model	Understand	Observe/ data mininig	Point of view	Ideate	Prototype/ modeling experiments	Test/ proof of value	Realize/ predict/ improve

added to the following table for a stronger illustration. The Design Thinking process for creative problem solving has been originally developed by Larry Leifer, David Kelley, and Terry Winograd at Stanford University. The aim of the approach is to bring together as many as possible different experiences and perspectives in respect to a possible problem situation. This approach leads to breakthrough innovations (Table 3).

The hybrid model has been designed to address a large part of the needs from both, Data Science and Design Thinking. The needs in Data Science for a common process model are primarily methods and tools to gain ideas and insights that cannot be obtained from data. Further, there is the need in Data Science to verify the correlations gained from the diverse data points or to develop new hypotheses together with experts from ethnographic research. The combination of Design Thinking and Big Data Analytics is also welcome in the communication of solutions (storytelling), which allow to transport emotions as well as hard facts. In multidisciplinary collaboration of mixed teams, one of the biggest wishes is to orientate oneself on a model that is easy to accept. For Design Thinkers it is of central importance not to lose sight of the most important element in Design Thinking, the customer/user with his needs, because it is one of the central starting points in the Design Thinking mindset. In addition, there is the need to continue using the tools used in Design Thinking and at the same time to open the existing mindset and enrich it with the requirements of more quantitative methods. For both, Design Thinkers and Data Scientists, it is of paramount importance to work together in multidisciplinary teams, to apply an iterative approach and to work with a common mindset in the problem and solution space. A common toolbox is described as optimal from both the experts in data and design.

The goal of the hybrid process is to establish a common approach to Data Science and Design Thinking in which teams can work together from problem definition to final solution and meet at defined times to exchange insights, discuss test results, and discuss the next experiments and procedures.

The Hybrid Toolkit

To achieve the best results, it is also necessary to work with a combined toolbox that incorporates the well-known tools and methods from Design Thinking as well as the tools and tools from Data Sciences. Fig. 3 shows a small selection of tools and methods that can be used in each phase. In Design Thinking, 100 additional tools can easily be added here and the list of models and tools in the field of Data Science is not shorter.

The tools from Design Thinking and Data Analytics are used across the entire application. For example, in the first phases of the Design Thinking process "Observe" and "Understand" Design Thinking tools are applied as described in "The Design Thinking Toolbox" (Lewrick et al. 2020). Typical Design Thinking Tools in these first phases are: Interview for Empathy, Explorative Interviews, Ask 5xWhy, Personas, Jobs-to-be-done up to Empathy Maps. Big Data Analytics allows you to collect and evaluate additional user data. On the one hand, this allows new insights to be gained, and on the other hand, insights from Design Thinking can be validated. User data is obtained from data sources ranging from social media and website analytics to data from neuroscience. Ultimately, it is a matter of selecting the appropriate tools to best understand the customer/user. This combination seems particularly important from a Data Science perspective, since a data-driven approach does not usually pay much attention to the activities involved in building empathy with the customer/user. Similarly, a hybrid toolbox can be used in the Ideate, Prototype, and Test phases. Typical Design Thinking Tools here are Brainstorming, Prototype to Test, Solution Interviews up to A/B Testing. The latter, for example, can be transferred directly and evaluated via data analytics, provided that two variants are tested and tried out via digital channels.

The same applies to all kinds of multivariant tests, testing emotional responses with facial recognition, and generating eye-tracking heat-maps with thousands of users in seconds. The entire spectrum of technologies significantly increases the speed to process data and help to emerge actionable insights much faster. Especially in cases where the behavior model is explored, such experiments can be valuable. Typical applications are, e.g., Willingness to pay analyses in the context of Minimum Viable Products (MVPs) or for testing individual features. In addition, the first data acquisition pipeline and analysis components can be developed on this basis, which later help to continuously improve the products and services at a later point in time. Big Data Analytics, in combination with Design Thinking, makes it possible to interpret user feedback and compare it with one's own observations better and faster.

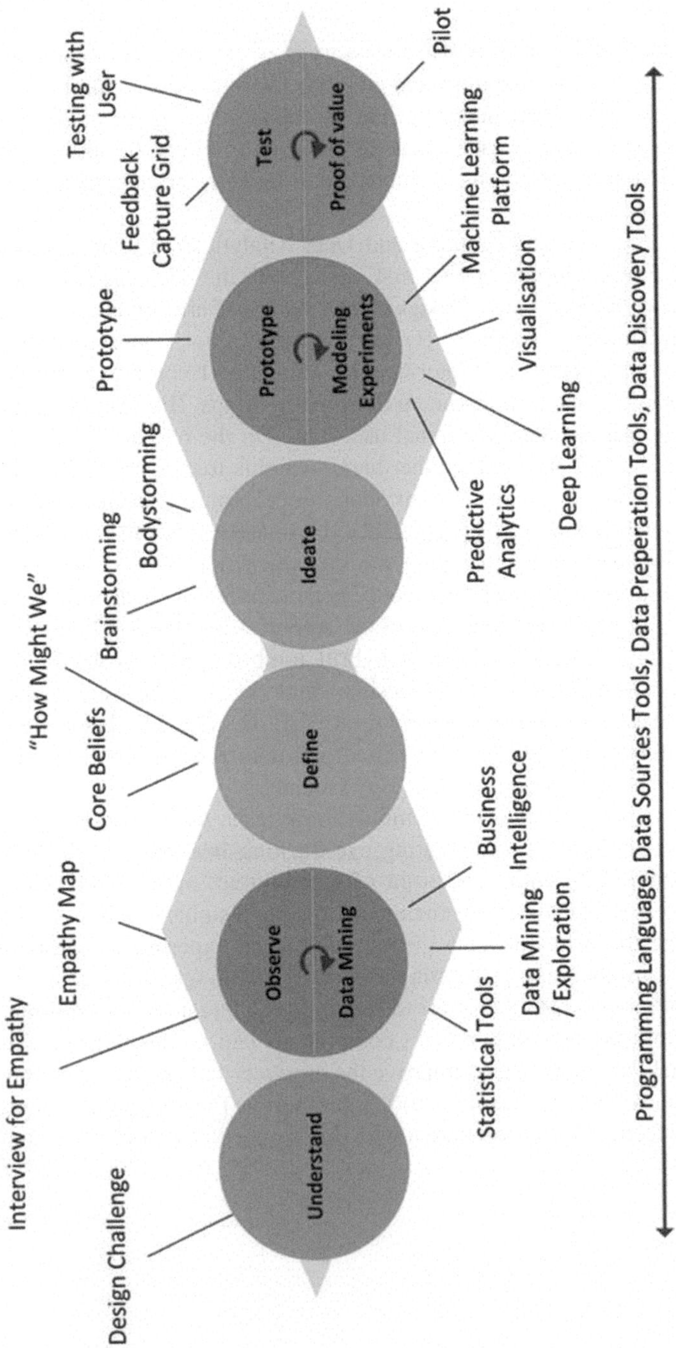

Fig. 3 The hybrid toolkit

The Power of AI-Enhanced Data Processing

With Design Thinking and AI hybrid teams can build a 720° real-time view of customers/users to determine their needs and customize services and products effectively. A 720° perspective on the customer refers to a three-dimensional understanding of customers based on big data analytics. It includes information about each customer's influencing capabilities, buying behavior, patterns, and needs. A 720° view enables enterprises to offer relevant products and experiences and predict future behavior. When properly implemented, this concept supports enterprises to leverage emerging technologies, social media and cloud-based services, and analytics to sustain lifelong and engaging customer relationships. This kind of insight-driven targeting is going to be essential in providing mass-customized services to customers/users. Machine learning and AI help to profile and cluster customers/users into ever-finer microsegments and even "segments of one." This approach enables a new dimension of hyper-personalization that can be used for experience-centric operations and self-service, customized services and products, targeted marketing and much more.

AI can also function as an assistant for hybrid teams, performing some heavy lifting with data collection and analysis. This includes finding patterns, making connections, and drawing conclusions. In addtion, AI supports the teams scrutinizing through numerous of surveys, interviews, observations, audio recordings, videos, and other user research data. Based on all the customer/user data it receives, AI can even predict what design pattern will work best. Further component libraries, style guides, and complex business ecosystem design systems can be generated in minutes. AI tools generate the design systems with ready-to-use code, simulation and apply updates automatically to the entire hybrid teams. AI is meant to work with creativity, not replace it—it is an important distinction.

On the opposite Design Thinking can be the basis for coding artificial intelligence, because it will not reach its full potential unless good designers guide the algorithm execution. Using Design Thinking to generate novel, user-centered applications for machine learning and define better parameters needed for the machine to execute upon. Both examples show applications of Data Science and Design Thinking beyond the hybrid model and at the same time show how data and design are in symbiosis rather than in competition.

Value Add of the Hybrid Model

Design Thinking has become a well-known problem-solving mindset promising customer-led solution through empathy and deep customer insights. It is implemented through an iterative process in multidisciplinary teams. Data Science aims to find patterns in growing data (Big Data), to generate stories from deep learning. Over the last decade both Design Thinking and Big Data Analytics have found its way into enterprises from all industries and both approaches are taught at almost all universities. In the meanwhile, also large Management and IT Consultancy powerhouses have explored and validated the value add of hybrid models.

The McKinsey study "Business Value of Design" (2018) and the Accenture study "How AI Boosts Industry Profits Innovation" (2017) released in separate research that shows how organizations that adopt AI and Design Thinking out-perform those that do not. As of today, however, design thinkers and data scientists still tend to work separately from each other in large organizations, and universities also tend to think separately from each other. The hybrid management model with a higher convergence between Design Thinking and Big Data Analytics aims to connect those two worlds. Using synergies and promoting collaboration should increase the efficiency and effectiveness throughout the whole process from problem to solution. Both Design Thinking and Data Science discover—versus define—the criteria for success; Data Science discovers the criteria for success buried in the data (patterns, codifying trends, and relationships) while Design Thinking discovers the criteria for success buried in the human interactions (using personas, empathy maps, and storytelling). What it needs is an open culture for sharing and collaboration, allowing all ideas to be worthy of consideration not mattering if the insights or ideas are created data-driven or customer-led. Organizations and universities must create a new joint learning culture through experimentation (and failure), and the willingness of culture to unlearn old methods that have been held as the gospel truth. Successful organizations and learning programs apply the hybrid model consisting of four central parts: The Hybrid process, the hybrid mindset, the hybrid teams, and applying a combined toolkit.

Practical Example

There are numerous examples of how the hybrid model has given rise to new ideas and innovation in recent years. Good practical examples can be found among the large business ecosystem initiatives, it becomes clear that the respective value propositions have evolved from a combination of Design Thinking and Systems Thinking, and that the offerings have been expanded through Big Data Analytics and AI, which now provide personalized services to customers. Ping An Insurance (Group) Company of China is a good example, which has also used in "Design Thinking for business growth" book authored by Lewrick (2021) for explaining the strong link between applying Design Thinking and new and emerging technologies. Ping An has become a world-leading customer and technology-powered retail financial services group. With over 214 million retail customers and nearly 579 million Internet users, Ping An is one of the largest financial services companies in the world. With AI, Big Data Analytics, and cloud computing, Ping An's financial offering has become more competitive, growing steadily while reducing costs, increasing efficiency, improving risk management and most importantly, enhancing the customer experience.

But the success of Ping An is not only based on the deployment of modern technologies. In the first place there are the new customer needs and in case of business ecosystem design the need for complementary offerings that make it easier for customers to obtain comprehensive solutions to their problems. Ping An realized by exploring the problem space that its customers wanted not only insurance but also a means of addressing their medical and well-being needs, it created Good Doctor. The Good Doctor platform offers 24-7 one-stop health care services that are provided by pharmacies, hospitals, and about 10,000 doctors. In the meanwhile, Good Doctor provides a portfolio on services to more than 65 million customers on a monthly basis. Most of the successful business ecosystem orchestrators offer their customers greater choice, even when that entails featuring competing offers. Ping An has become a good example of applying Design Thinking for business growth with a strong focus on the data ecosystem which allows to expand value propositions constantly. On top Ping An has realized that the mindset and collaboration of different teams become key to success.

Top Leaders at Ping An are called upon to promote the cooperation of interdisciplinary teams even more strongly in the future in order to promote innovations that require the cooperation of different teams.

On the Point

As briefly described in this contribution Data Science can be used very well in combination with Design Thinking. The hybrid model aims to not only enrich Design Thinking with Data Science, but to establish from the beginning a common mindset in which different T-shaped team members contribute their skills throughout the whole process, in order to get better solutions in a faster and more effective way.

To sum-up: The combination of several data sources and insights (qualitative & quantitative) leads to:

- A better understanding of the customers/users and his needs
- Improved decision confidence
- Validation and alignment of insights
- Profound decision making and validation of assumptions
- Management approval of solutions enhanced by quantifying intuitions
- Contextualization of data in order to link data insights with stories
- A faster iterative problem-solving process and mass customization
- Reducing risks.

The application of the hybrid model and AI has already and will even more changing the way we work. The hybrid mindset, tools, and process will make every team member a better problem-solver and bringing better and more personalized products and services to customers/users. The practical example from Ping An has shown how Design Thinking, Systems Thinking and the combination with Big Data

Analytics led to superior business ecosystems in which complementary offerings are designed around the new or changed needs.

References

Lewrick M (2021) Design thinking for business growth: how to design and scale business models and business ecosystems. Wiley, New York

Lewrick M, Link P, Leifer L (2018) The design thinking playbook: mindful digital transformation of teams, products, services, businesses and ecosystems. Wiley, New York

Lewrick M, Link P, Leifer L (2020) The design thinking toolbox: a guide to mastering the most popular and valuable innovation methods. Wiley, New York

The Collective Process Framework DTScrum for Integrating Design Thinking into Scrum

Daniel Gadner and Michael Felderer

Introduction

It is a well-known fact that the accelerating pace of change in today's digital world comes along with disruptive technologies, volatile requirements, wicked problems, and sophisticated demand. This fast-paced environment is often referred to as the time of digitalization, where humankind must inherently transform the traditional ways of product development practices, internal organizing logic, and customer interactions to stay competitive. In order to describe this transition, Gray and Rumpe (2015) use the business-oriented definition of digitalization from Gartner: "Digitalization is the use of digital technologies to change a business model and provide new revenue and value-producing opportunities; it is the process of moving to a digital business." (p. 1319). Despite the benefits that digitalization brings along, it is difficult to maintain the course in the ever-increasing flood of data resulting in unpredictable situations and complex problems (Gray and Rumpe 2015). Hand in hand with such digital transformation efforts goes the diffusion of digital technologies, and therewith, the digital disruption. Technological diffusion is known for its far-reaching effects, as, for example, the rapid pace of change in market needs (Abrell et al. 2016). Stoneman and Battisti (2010) define the technological diffusion "[...] as the process by which the market for a new technology changes over time and from which production and usage patterns of new products and production processes result." (p. 733). In this context, firms must learn to adapt to a volatile environment and react quickly to change. The challenge is that today's customers are

D. Gadner (✉)
Internetstores, Stuttgart, Germany

M. Felderer
University of Innsbruck, Innsbruck, Austria

Blekinge Institute of Technology, Karlskrona, Sweden
e-mail: michael.felderer@uibk.ac.at

J. Hehn et al. (eds.), *Design Thinking for Software Engineering*, Progress in IS,
https://doi.org/10.1007/978-3-030-90594-1_5

very well-informed about the capabilities of new technology and are consequently more demanding. Competition is hard as customer needs evolve quickly. It is getting difficult for firms to anticipate the forces of change. The corporate environment is becoming confusing (Fahey 2016). As a result, the timely reaction to changing requirements is more important than ever. It can be best managed by using agile frameworks for digital product development (Fonseca and Domingues 2017). Those often come along with practices that enable quick accommodation to change rather than following a defined plan. Especially in the software engineering domain, agile means short development cycles resulting in fast reactions to changing customer needs (Abrell et al. 2016). Though many companies are jumping on the bandwagon of agile software development practices with their fast feedback loops and flexible functioning, they must not forget, resulting in complex problems. The so-called wicked problems are one of the reasons why firms fail to survive in the time of digitalization. Here, tackling such a class of problems plays to the strength of Design Thinking. The background section will cover the functioning of it, as well as the principles of the Scrum framework. In addition, potential obstacles and challenges of an integration will be exemplified. Based on input retrieved from already conducted expert interviews and practical observations, a conceptual model was developed to overcome beforehand identified challenges. This integrative model is called Collective Process Framework and starts with the Multidisciplinary Knowledge Café, diverging into two process areas running in parallel. The term "Collective" is here used as it describes the acting in the context of software development processes in a cooperative way. In other words, the Design Thinking process, as well as the Scrum process, were emblematic put next to each other to produce a product of collective effort. This means, one area focuses on irrational beliefs and lateral thinking, whereas the other one sticks to causal reasoning and organized thinking. In this context, Plattner et al. (2012) mentioned that "[. . .] Design Thinking focuses on those 'fuzzy' aspects of a design problem, which are in purely engineering-led approaches left aside and is thus suggested as a useful supplement to a problem perception and solving in 'traditional' IT development approaches." (p.231). Although, one might argue about the contradicting focus each process pursues on its own. On the one hand, Scrum has a focus on iterative development more likely to respond to change rather than planning for it. On the other hand, Design Thinking has the goal to discover change opportunities that solve customer problems and thus imply movement towards an idealized state. This situation provides an additional motivation to integrate Scrum and Design Thinking in this contribution to benefit from synergies and minimize potential problems.

This contribution is structured as follows. We first present the necessary background on Scrum and Design Thinking. Then, we discuss obstacles and challenges of the integration of Scrum and Design Thinking. As main contribution, we present our collective process framework DTScrum. Finally, we conclude this contribution.

Background

The central frameworks, Scrum and Design Thinking, are both very helpful when it comes to complex problem-solving in a dynamic environment. Scrum, on the one hand, can be named as an effective approach to developing customer-focused solutions. That is because, referring to Vetterli et al. (2013), the main task of Scrum is to implement the user stories which represent the customer requirements in clear form. On the other hand, Design Thinking can be named as a useful process framework to identify real customer problems, redefine complex problems, challenge assumptions and develop novel product designs. Plattner et al. (2012) mentioned that requirements, especially for software systems, should be collected and defined in collaboration with the user, often via the means of Design Thinking.

Scrum

Jeff Sutherland and Ken Schwaber coined the term "Scrum" in the context of agile software development in 1990 and declared that constant evaluation of requirements, plans, and results is the rudiment for a common and natural understanding of how to respond quickly to change. Moreover, Scrum practitioners should embrace change. It is better to respond to change rather than planning for it. All team members ought to discuss on daily basis topics such as what had been already delivered and what kind of tasks are still open. Hereby the Scrum framework promotes an open meeting culture which should assure constant and effective communication inside the team and between all involved parties. This also contributes to a common understanding about the same line of vision on what to achieve, which is mandatory for success (Nachbagauer and Ortner 2015). Next, the scope is not fixed. More the scope is initially mirrored in kind of a project plan, which is called the product backlog. It can evolve and can be seen as a living artifact. Scrum practitioners must always keep in mind that the next set of features to work on could change because of the unpredictable customer and market behavior. Therefore, the product backlog should be updated with each completion of a development cycle. Such a cycle is called an iteration or sprint and describes the time frame during which the team completes predefined user stories. Often, it is depicted as a circle because there is not only one but many iterations. Part of each sprint is a daily meeting, called daily Scrum. As mentioned before, this should foster a communicative culture within the Scrum team (Schwaber 2004). The result of each sprint is called an increment of functionality. Here, the whole team is responsible for the fast and agreeable delivery of the software product. Working functionality is delivered to the customer as soon as possible in terms of small and frequent repetitions (Dzamashvili Fogelström et al. 2010). After each sprint, the team should always look back on how they reached the actual state of delivery and analyze its way of working. All in all, Scrum can be named an iterative, incremental process. Or in other words, Scrum is an agile project

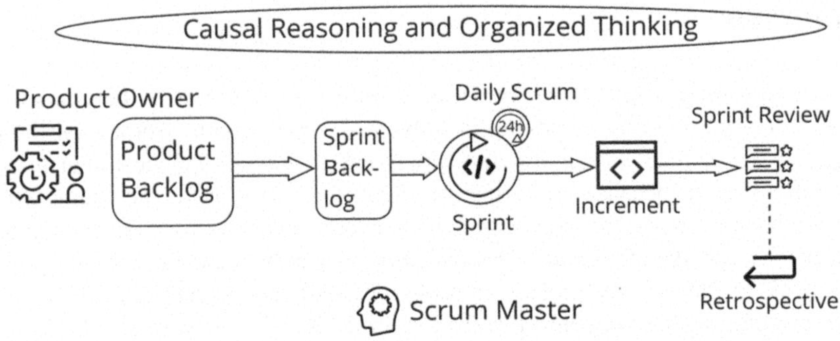

Fig. 1 Scrum process area

management framework mainly used for software development projects (see Fig. 1 for an overview of the Scrum process), which includes change embracing properties (Schwaber 2004).

Design Thinking

In accordance with the research of Gerstbach (2016), it can be stated that most of today's people associate the physical figure of a product with a successful design. Even though successful design evolves out of meaningful and effective development of products incorporating the needs and wishes of users. Hereby the main task of a designer must be the investigation of customer behavior in order to solve the right problem (Gerstbach 2016). In respect to the Collective Process Framework, Damien Newman's illustration of the design process is used to exemplify the characteristics of the Design Thinking process (see Fig. 2 for an overview of the Design Thinking process). The look-alike scribble is called design squiggle and represents the way from research over developing prototypes to finally establish a clear design. Furthermore, it shows that the way to the desired solution might be uncertain in the beginning, but in the end, it will evolve into a single point of clarity. In other words, every solution is born out of uncertainty (Naiman 2017). If necessary, one might even need to take a step back or start all over again; because failure is knowledge and knowledge means success. The path to a successful design consequently unfolds in iterative loops moving back and forth across phases. "With this acknowledgment, it is perhaps also worth noting that whether the process is intentional or not, it will assume a significance upon the final design outcome" (Schneider et al. 2013, p. 117). As we know by now that there is no ultimate single solution to a problem, four factors of success, each individually weighted dependent on the respective situation will be described now. At first, everybody's own intuition is decisive. Due to the explorative character of the Design Thinking process, undiscovered problems and opportunities are more likely to be revealed by trusting the gut instinct. Furthermore, no generated idea is nonsense. Every unique way to a solution should be chased.

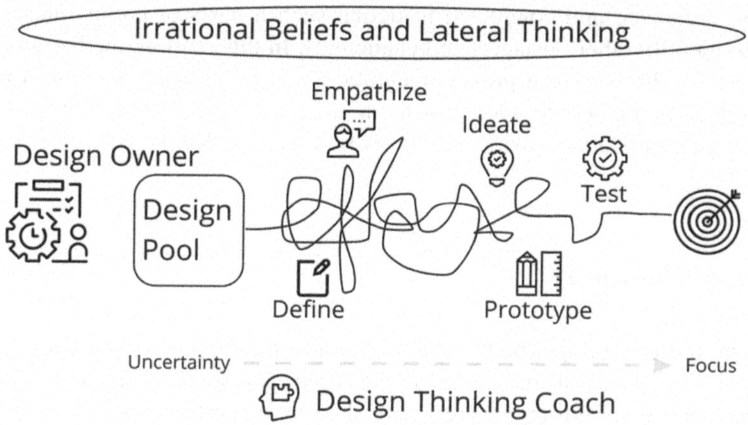

Fig. 2 Design thinking process area

That leads already to the second property, which is the trial-and-error principle. Besides the negative aspect of trial and error, which is nearly impossible to predict the duration of the ideation phase, this principle is a great way to gain knowledge. There are many ways to solve a wicked problem. Especially prototyping should be used to reveal the right problem and possible solutions to it. All ideas obtain the opportunity to be sufficiently tested and improved till one idea stands out from the others. Resulting the third factor of success is the direct implementation of the ideas in the form of prototypes. Now the last property is necessary to identify the limits of the human-centered design approach. Those boundaries should not be exceeded. Within the defined scope Design Thinking can unfold its comprehensive capabilities to innovate. The boundaries are congruent with the desirability, viability, and feasibility dimension. As a result, the appropriate balancing between the feasibility of the technology, the viable business dimension, and the desired customer needs is essential. Dependent on the situation every aspect should be given more weight, sometimes less weight, always with the human being in mind (Gerstbach 2016).

Obstacles and Challenges of the Integration

The successful integration of a proper design understanding in the industry sector, marketing branch, or public had already been an important topic several years ago. However, little is yet known on how to make effective use of Design Thinking in the context of agile software development methods. Buchanan (1992), states hereof that, "Without appropriate reflection to help clarify the basis of communication among all the participants, there is little hope of understanding the foundations and value of Design Thinking in an increasingly complex technological culture." (p. 8). Thus, as a part of former research, interviews with experts in Design Thinking and/−or Scrum had been organized and conducted to define such a basis of communication and further to identify other obstacles and challenges. In this section, the barriers which may occur on the way to a proper integration will be pointed out, and a process framework showing how to overcome them, on a conceptual level, will be depicted next. For exemplification, own experiences will be included as well.

Resource Allocation

Of course, the first thing which might come into mind when mentioning possible challenges of a practical integration is the topic of resources. In most of today's companies, resource restrictions are a critical issue. It is pretty rare that anyone finds themselves in the fortunate situation of being able to focus 100% on a specific problem without any other distractions. Especially, that if all team members are committed to this one project to the full extent. As Design Thinking relies on the involvement of the user as well as the knowledge of people from different domains, it is often difficult to agree on resource capacities. So, depending on the ratio of wicked problems to more tangible problems, you may need to focus your resources either on one or another mindset. Design Thinking itself tries to obtain the user feedback and expertise of other professional roles as often as possible. Therefore, the resources must need to be allocated before the project starts, i.e., in the form of a sensible resource allocation meeting. This will be described later on in terms of the process framework, our starting point for any integrative Design Thinking—Scrum projects. Here, it is important to fine-tune the occupation of resources and the point in time when they are presumably available. So the emphasis on the work must be based on the degree of resource utilization in an appropriate ratio to discover complex problems and ordinary problems. Too many resources would have been occupied with several user research topics and ideation sessions before the actual development phase, in case of Design Thinking is used as an upstream process in front of the Scrum steps. According to this, another challenge is the limited budget for a human-centered design project, as well as the pressure from the management to develop something potentially shippable in a short period. It is, therefore, important that the management understands that agreeing to incorporate Design Thinking in software development projects is a comparatively more time-consuming commitment without guarantee for a successful product idea as an outcome. Instead, every insight gained by the problem discovery and definition phase is a small step toward understanding your customer demand to its full extent. Thus, the resource investment seems to be justifiable.

Competing Views and Different Kinds of Problems

According to the different kinds of obstacles to face, it is necessary to mention that despite all the similarities between Design Thinking and Scrum, they are very different in nature. The Design Thinking's problem-solving approach considers socially ambiguous aspects that build upon heuristics and situational reasoning, whereas the corresponding problem-solving patterns of Scrum rely more on the orthodox engineering design paradigms and analytical thinking (Plattner et al. 2012). With this thought in mind, we must be aware of the competing views between the design domain and the software engineering domain. The different purposes and distinct kinds of problems to be treated by the frameworks could be a reason for conflict when trying to collaborate with each other. Also, the oppositional mindset of the Design Thinking team and the more analytical thinking Scrum people can be seen as a challenge here. Consequently, the question arises of how to use two frameworks together when both have a diverse purpose. "Whereas Design Thinking allows dealing with the ambiguity of design problems as wicked problems, the thinking of IT engineers instead supports the effective technical realization." (Plattner et al. 2012, p. 230). Also, in practice, you can feel the tension between those opposed characters. The most problematic thing is when the more open-minded approach of Designers clashes with more analytical-oriented Scrum practitioners. Both somehow feel a thread in the counterpart's way of thinking, as it might disturb its own way of working. Scrum practitioners might feel held back by the Design Thinking counterpart as the Design Thinking people might feel annoyed by the disbelief of developers in the value creation of the Design Thinking steps. Here, it is necessary that all team members should be sensitized to the opposing mindset. Every project member should have an open mind toward new thinking and working mode. However, the Scrum framework's capability to support the path of understanding the problem and, in turn, defining certain requirements is limited. Approaches, like Scrum, are not yet able to cope with a class of problems demanding techniques reaching from further exploring problems to fully understand them. Especially when thinking of problems occurring in the fast-paced technological world, the characteristics of complex problems, especially in the context of the human-centered design approach, are of high relevance. Especially today's software development projects demand more and more for human-centered design approaches to tackle what is often referred to as wicked problems, i.e., unknown and inherently volatile requirements. Nevertheless, identifying requirements for a solution to tackle the right problem is a very sophisticated process. This might also be the reason why firms fail to survive in the time of digitalization. That is to say, while software development approaches aim at transferring customer needs into rapid development cycles as a means to develop software products iteratively and incrementally, we often still need to first shift our attention to framing and changing the actual problem. To achieve this, Design Thinking should be used to challenge assumptions, redefine complex problems, and even explore new possibilities and paths for problem solving. On the other hand, Scrum is a bit more dedicated to rapidly develop solutions for tamed or ordinary problems, based on user stories that represent the customer requirements in clear form. For example, if you have a problem that is very well defined, you would not necessarily need to conduct a full-blown Design Thinking project, as it would be too big of an investment and a waste of resources. This is because here, you already know about the problem space and the resulting customer pain to be solved. For example, in web development, it is already common knowledge that an "add-to cart" button on a product detail page must be shown on the first view and designed in a prominent way to boost the conversion. Here you can benefit from already existing knowledge and do not necessarily need to start from scratch. So, in this case, one can directly start with the development sprint.

Coordination and Communication

Important for this kind of collaborative procedure of using Design Thinking and Scrum together is constant and regular communication and coordination between all participants. This also involves stakeholder management. Especially in terms of Design Thinking, with all its gatherings of people with diverse backgrounds also the different types of personalities working together must be considered. This can create a conflict between the participants, which, at worst, could also have a bad influence on future cooperation. In order to prevent this potential conflict, an emphasis must be put on effective communication. Nevertheless, a balance between meeting overload and too little communication, which might result in knowledge silos, is crucial. It is important to never underestimate the power of gathering. As soon as several people meet to work together on a concrete topic, problems can be defined, and ideas to solve them are collected. Here, the ones responsible for the output of any implementation must be informed about the current state. People who will be working on the solution to the problem in the near future do not only want to receive a list of requirements to be worked on. Rather they prefer to be part of the journey from the very beginning. Otherwise, people might feel excluded and think they missed their chance to contribute their thoughts and ideas to it. Sometimes it is just the simple human behavior of being affronted after somebody has told them what and how to do something, which puts a successful Design Thinking Scrum project, or any project in general, at risk. This can also result in a lack of motivation and decreases the commitment to the project. When it comes to organizational structure, hierarchical constraints must be mentioned as a possible factor that can also decrease the effectiveness of applying a design methodology such as Design Thinking in the software development domain. An example of something which restricts creative work is design standards given by management. Therefore, "[. . .] a balance needs to be found between corporate requirements and creative freedom" (Häger et al. 2015, p. 264). Further, it is crucial that everybody being part of interdisciplinary teams should know their role and associated responsibilities within the project. It is a completely different situation working together with someone on a regular basis, knowing the team's internal structure and hierarchies, in comparison to someone who has just recently been added/begun working with it. This can create tension, especially when it comes to an overlap of responsibilities. That is also why ownership is a very important part of this topic. If you are responsible for a specific piece of the project, you are less likely to feel threatened by other one's actions to affect your work.

The Collective Process Framework DTScrum

Especially today's software development projects demand more and more for human-centered design approaches to tackle what is often referred to as wicked problems, i.e., unknown and inherently volatile requirements. To achieve this, Design Thinking should be used to challenge assumptions, redefine complex problems, and even explore new possibilities and paths for problem solving. Now, as there are already several kinds of research work that agree on the idea of putting a Design Thinking approach together with Scrum, we are tackling the issue of really synthesizing those different mindsets and working modes. In this sense, the conceptual model which we intentionally describe as a *Collective Process Framework* should be understood as a visual roadmap of:

- Artifacts
- Roles and responsibilities
- Activities

Toward the development of a software-intensive solution using both Design Thinking and Scrum. The model is called the collective process framework, with reference to the work by Malone et al. (2009), who investigated the effectiveness of individuals working together in a collaborative environment. Collective here means that both, Scrum and Design Thinking, are functioning on their own but are connected in several ways throughout the overall solution discovery and development process in order to profit from each other's respective strengths. Therefore, the collective process framework emphasizes a collaborative working mode and mindset. This is achieved by putting the Design Thinking phases in parallel to the Scrum framework, whereby feedback cycles and coordination meetings are promoted in order to maintain a constant connection between the Design Thinking work and the Scrum work. Based on Malone et al. (2009), "The phrase we find most useful is collective intelligence, defined very broadly as groups of individuals doing things collectively that seem intelligent." (p. 2). Consequently, the term "Collective" should, on the one hand, be used to describe the multidisciplinary characteristic of the process framework, on the other hand, to describe the in parallel functioning Design Thinking and Scrum processes. In other words, the Design Thinking process, as well as the Scrum process, were emblematic put next to each other to produce a product of collective effort. Now having a closer look at the conceptual model, the spectator will notice so-called process areas, currently four in number. The following represents the skeleton of the model:

- Multidisciplinary Knowledge Café
- Design Thinking process area
- Scrum process area
- Product Backlog Design Matching

Multidisciplinary Knowledge Café

The first area must be seen as the groundwork for a successful design implementation approach and, therefore, starts with a multidisciplinary knowledge café that involves:

- Bringing people with different backgrounds together
- Connecting diverse points of views
- Preparing a rough roadmap in order to plan and allocate resources appropriately upfront
- Identifying and classifying problems

So, the term multidisciplinary here is used in reference to the Design Thinking mindset, which emphasizes knowledge assimilation regarding putting different points of view together. Incorporating people with diverse fields of expertise is an advantage. Here, Buchanan used the term liberal arts to describe a vision of an encyclopedic education of various natural sciences and mathematics, philosophy, and social sciences. As a result, Design Thinking in the twentieth century must be recognized as the new liberal art of technological culture. Design is an integrative discipline to complement art and sciences (Buchanan 1992). In this case, the advocates of the new liberal art say that "[. . .] the proper study of mankind is the science of design, not only as of the professional component of a technical education but as a core discipline for every liberally educated person." (Simon 2019, p. 138). This is first, effective in a way that every aspect of a problem is mainly covered, and second, it will be talked about any concerns and ideas regarding future work. Thus, such a creative and open-minded approach results either in the identification of a concrete problem set or in the revelation that there, in fact, is no real problem, which can save a considerable amount of time. Trying to solve something which you are not really sure about exist is a complete waste of time. Besides the basic principles of the Multidisciplinary Knowledge Café, namely fine-tuning resource capabilities, connecting diverse perspectives, sharing collective discoveries, and acknowledging different opinions, its main purpose is the identification of subproblems and their classification. An emphasis must be put on the implication that it will be distinguished between wicked problems and well-known, or "ordinary" problems. The separation into different kinds of subproblem is important as the next process areas of the collective process framework have different capabilities to handle the outputs of the Multidisciplinary Knowledge Café. In short, it will be distinguished between ordinary problems and wicked problems. For example, the purpose of Design Thinking is directed towards complex problem solving and novel idea generation, whereas the Scrum process often deals with ordinary or ill-defined problems. Thus, the Design Thinking part deals with the treatment of wicked problems and innovation opportunities, collected in the so-called "Design Pool" which is the counterpart of the Scrum processes product backlog. Wicked problems are real-world problems that are difficult or impossible to solve for some reason. There are no solutions in the sense of definitive and objective answers. One could say, wicked problems refer equally to problems of design and planning. The scrum approach is a bit more dedicated to rapidly develop a solution for tamed or "ordinary" problems. This class of problems is known for its well-defined requirements where the solution is clear. Known algorithms can be applied straight to come up with the solution. Even though the wording might imply it, "ordinary" problems are not less worthy of being solved than wicked ones. But it would not be sufficient to re-discover the same problem again and again. Therefore, it is important to differentiate between those. For example, if you have a problem that is very well defined, you would not necessarily need to conduct a full-blown Design Thinking project, as it would be too big of an investment and kind of a waste of resources. This is because here you already know about the problem and the resulting customer pain. For example, in web development, it is already commonly accepted that the basic structure of any product detail page includes an image of the product, the title, the price, variation selection, and the add-to-cart button. Maybe in the past, a problem existed, that customers have not really had added products to the cart. Now we know that it is decisive to show a clickable element, also known as the "add-to-cart" button, on the first view as it is the main element on such a template to boost the conversion. In this case, you can benefit from already existing knowledge and do not necessarily need to start from scratch. Resulting, one can directly start with the implementation sprint. To make use of both, the collective process framework is not really about integrating one framework into another. It is about finding ways of planning both together because they can also work on their own very well. An emphasis must be put on a collaborative working mode and mindset. This is achieved by putting the Design Thinking phases in parallel to the Scrum framework, whereby feedback cycles and coordination meetings are promoted in order to maintain a constant connection between the Design Thinking work and the Scrum work. Such collective functioning is vital for the ultimate implementation of a successful product design. It is depicted in the conceptual model as process areas two and three, which will be described now.

Diverge into Design Thinking and Scrum

The class of problems must be structured, categorized, and prioritized. Therefore, the second and third process areas start with backlog management. Here it is important to get a better overview of the work to be done. The different class of problems which have been discussed beforehand are going to be revised. On the one hand, there is the Scrum framework, focusing on organized thinking and causal reasoning, and on the other hand, there is the Design Thinking approach, more of lateral nature. First, coming to the Scrum process, which stands for the principles of cause and effect. It is all about thinking which event leads to one another. Especially, communication within distributed systems and its determination of the sequence when several instructions should take place is crucial. So, this way of thinking leads to a causal order in which the output of one action leads to and, or is needed by the following operation. As a result, processes can be planned in a more holistic way. It emphasizes analyzing and categorizing conceptual information using a systematic and logical thought process. In comparison to this incremental way of thinking, the Design Thinking approach is more about the opposite. Here, lateral thinking stands for analysis by intuition. Mental leaps are no shame, and not every interim finding must make sense in the first place. Especially odd ideas are welcomed. Further, initial situation and boundary conditions must not be seen as immutable, rather as an opportunity to shape the context for its own purpose to open up new perspectives. You should force yourself to believe in something which you normally would not believe due to objective evidence retained by intrapersonal cognitive structures. In this sense, Design Thinking ends with a good understanding and confidence of ideas.

Irrational Beliefs and Lateral Thinking

Process area two is showing the before-mentioned illustration of Damien Newman's design squiggle, which refers to the design process and it is way from research over developing prototypes to finally establish the clear design (Naiman 2017). Here, irrational beliefs and lateral thinking minimize the risk of developing a solution towards a non-existing problem and uncertainty about the right problem by engaging customers through a series of prototypes to learn, test, and refine concepts. Learning from failures of previous iterations is essential. Now, the roles corresponding to the Design Thinking work within this collective process framework will be explained. Here, the role of the Design Owner is the counterpart of the Product Owner and, as a result, responsible for the Design Outcome, for instance, the mock-up. Important to mention here is that the Design Owner decides with the consultation of the Design Thinking team which elements of the Design Pool should be part of the next design step and if going back or forth to any other step might be beneficial. Another option would be an innovation opportunity that seems profitable so that more investigation must be made. Either way, the Design Owner triggers the design sprint, which in turn will be run iteratively by the Design Thinking team, including the Design Thinking Coach, and in some cases, subject matter experts. The Design Thinking Coach can be seen as the counterpart to the Scrum master.

Causal Reasoning and Organized Thinking

For the correct application of Scrum, the applicant must remember the major principles of agile software development. First, the development of features should happen right away. Working functionality is delivered to the customer as soon as possible in terms of small and frequent repetitions. Additionally, embracing change is a crucial factor. Therefore, Scrum practitioners should be aware that it is better to respond to change rather than plan for it. Next, the scope is not fixed. More the scope is initially mirrored in kind of a project plan, which is called the product backlog, which is one of the four key building blocks of the Scrum process. It can evolve and be updated between the completion of each iteration. It is kind of a learning process. The other elements are the increment of functionality, the 24-hour inspection meeting, and the time frame of an iteration cycle called a sprint. As explained yet, Scrum is an iterative, incremental process. It all starts with the product backlog. This artifact is a list of requirements that are mostly requested by the customers and, thus, by the Product Owner. An iteration of development activities is represented as a circle because there is not only one iteration but many, one after another. The result of an iteration is the potentially shippable product increment. So, if a final increment can be presented to the customer, everything starts from the beginning. The problem with Scrum is that most organizations nowadays do not allow the direct collaboration of users and the development team. The principles of Design Thinking could help the Product Owner or any other role involved in the product development process to establish a better understanding of how to tame complex problems, foster innovative thinking, and have a customer-focused mindset.

Converge

An important success factor is the same line of vision, which in turn is crucial to synthesize all tasks. Therefore, it is not enough to just let the two processes run in parallel in order to benefit from a collective process approach. It must be ensured that both frameworks benefit from the advantages of their counterparts and still utilize their own strengths. Important for such a collaborative procedure is constant communication and coordination with each other. That is why we want to introduce the fourth process area. In this context, the Design Thinking output, like, for example, nonfunctional prototypes, should be used as an input for further development cycles. This means that the insights about customer needs and tamed problems must be included in the development sprints of the Scrum team. "Ideally a requirements specification could include a description of basic assumptions, needs, and knowledge of the problem domain, needed for designing and implementing an information system." (Bubenko 1995, p. 160). Now, it is all about bringing the work back and learning from each other's achievements. Important for such a collaborative procedure is constant communication and coordination with each other. The

Multidisciplinary Knowledge Café, which brought together different professions and personalities in the first run, can also be used to converge the beforehand diverged working teams. So, for this converging step, an artifact such as a backlog can be used again as a merging tool to bring the people back together and collect the different achievements, ideas, and problems in a shared area. This meeting is called "Product Backlog Design Matching" and offers attendees the possibility to discuss newly developed functionality, improved mock-ups, tamed problems, and clearly defined innovation opportunities. Further, recently appeared complex problems that cannot be implemented due to non-existing clear definition are part of this pool. The attendees should feel like to be cushioned against the "daily" work (reality). This melting pot is the heart of past work and the foundation for future achievements. This means that the insights about customer needs and tamed problems must be included in the development sprints of the Scrum team. It is important to hand over the knowledge of each other's tasks and achievements. When having a look at the parallel functioning processes on a holistic level, there is still the connection to the end-user. This is also where you get the knowledge on which new features and functions should be integrated into the next sprints to solve the right problem. Further, issues can be discussed so that misunderstandings can be excluded, and maybe another point of view will reveal an alternative way to solve an issue. The vision which has been developed in the Multidisciplinary Knowledge Café, therefore, can evolve over time by keeping the core problem and goal in mind. It will be noticed that the knowledge hand-over is a critical but also indispensable milestone during the solution development process. Critical, because there is so much knowledge that is implicitly within the Design Thinking team and its possible loss of achievements and insights due to the poor hand-over to the other or a completely new team, which might happen due to capacity or resource constraints. Mainly this is critical because Design Thinking does not document that much. And indispensable because ideas exist, which you need to implement to experience it and to see if the idea is working. In other words, the team is performing some Design Thinking stories. Already existing user stories on the part of the Scrum product backlog can be adapted or removed, and new user stories can be established. As the Product Backlog Design Matching finishes, the "daily" work begins again.

Conclusion

All in all, we presented the Collective Process Framework, which unifies Design Thinking and Scrum on a conceptual level. The framework, visualized in Fig. 3, comprises four process areas, and it should be understood as a visual roadmap of general process items, including:

- Artifacts, such as mock-ups and ideas and their relationship development artifacts such as the product backlog.

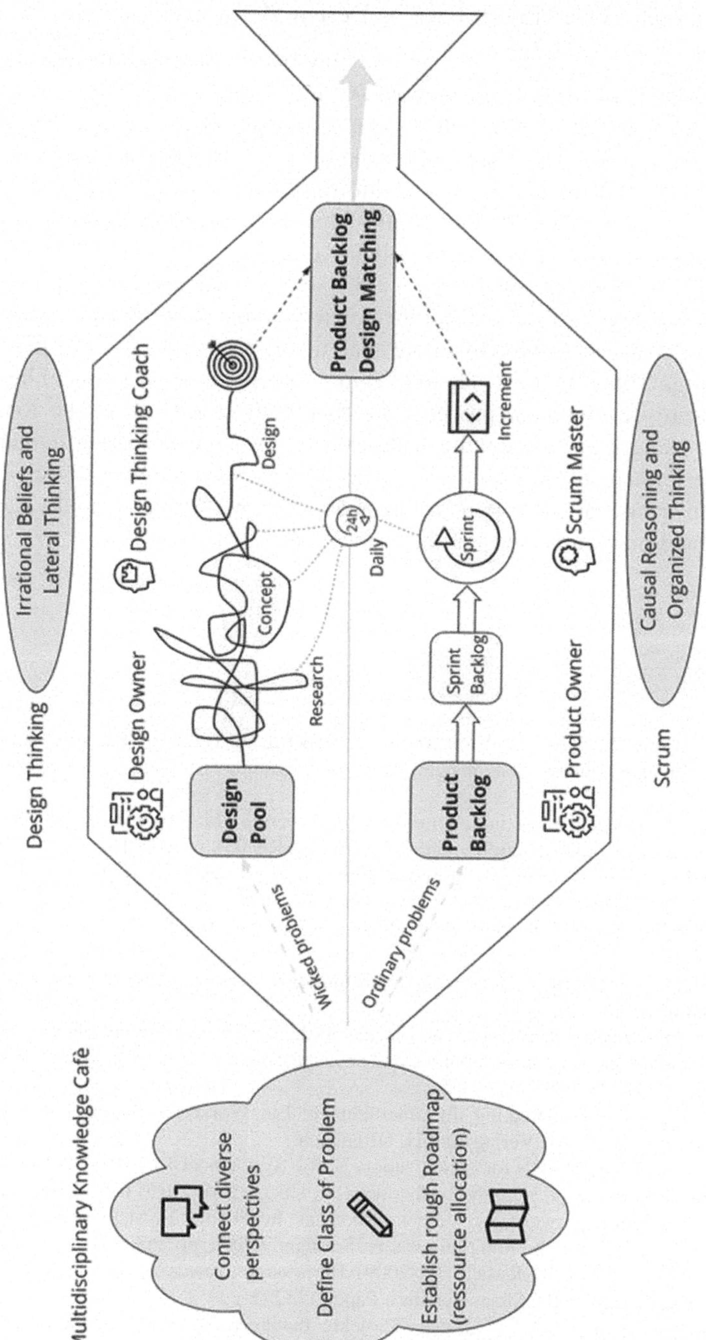

Fig. 3 Collective process framework

- Roles, such as the design owner building a counterpart to the Product Owner.
- Activities, such as the Product Backlog Design Matching.

This roadmap serves, as we argue, as one important step toward the development of a software-intensive solution using both DT and Scrum. Both frameworks benefit from each other and still utilize their own strengths. Here, Design Thinking is frequently identified as an engaging process and methodical framework for approaching complex problems in a multidisciplinary way. Such a design work often tames complex problems and results in novel design solutions. Scrum is a simple framework for effective team collaboration in a dynamic environment, in which communication and collaboration are essential elements. Further, it provides guidance for effective product development in a dynamic environment. Now, due to the lack of social aspects within the more analytically arranged software engineering domain, Design Thinking must help to integrate a proper understanding of humanity in the software development process. So, this shall contribute to the long-term objective of using Design Thinking within the Software Engineering domain. We consider the refinement of the process areas by empirical studies, especially case studies in an industrial context, as future work, and we cordially invite interested researchers and practitioners to join our endeavor.

References

Abrell T, Pihlajamaa M, Kanto L, Vom Brocke J, Uebernickel F (2016) The role of users and customers in digital innovation: insights from b2b manufacturing firms. Inf Manag 53(3): 324–335

Bubenko JA (1995) Challenges in requirements engineering. In: Proceedings of 1995 IEEE international symposium on requirements engineering (RE'95), pp 160–162

Buchanan R (1992) Wicked problems in design thinking. Des Issues 8(2):5–21

Dzamashvili Fogelström N, Gorschek T, Svahnberg M, Olsson P (2010) The impact of agile principles on market-driven software product development. J Softw Maint Evol Res Pract 22(1):53–80

Fahey L (2016) John c. camillus: discovering opportunities by exploring wicked problems. Strategy & Leadership

Fonseca LM, Domingues JP (2017) How to succeed in the digital age? Monitor the organizational context, identify risks and opportunities, and manage change effectively. Manag Mark Chall Knowl Soc 12(3):443–455

Gerstbach I (2016) Design Thinking im Unternehmen: Ein Workbook für die Einführung von Design Thinking. GABAL Verlag GmbH, Offenbach

Gray J, Rumpe B (2015) Models for digitalization. Softw Syst Model 14:1319–1320

Häger F, Kowark T, Krüger J, Vetterli C, Übernickel F, Uflacker M (2015) Dt@scrum: integrating design thinking with software development processes. In: Plattner H, Meinel C, Leifer L (eds) Design thinking research: building innovators. Springer, Cham, pp 263–289

Malone TW, Laubacher R, Dellarocas C (2009) Harnessing crowds: mapping the genome of collective intelligence. MIT Sloan Research Paper 4732(09)

Nachbagauer A, Ortner G (2015) Globale Projekte managen: Neue Wege für die weltweite Projektarbeit. Symposion Publishing, Düsseldorf

Naiman L (2017) Design Thinking as a strategy for innovation. The European Business Review. www.europeanbusinessreview.com/design-thinking-as-a-strategy-for-innovation. Accessed 16 Feb 2019

Plattner H, Meinel C, Leifer L (2012) Design thinking research. Springer-Verlag, Berlin Heidelberg

Schneider J, Stickdorn M, Bisset F, Andrews K, Lawrence A (2013) This is service design thinking. BIS Publishers

Schwaber K (2004) Agile project management with Scrum. Microsoft Press, Redmond

Simon HA (2019) The sciences of the artificial. MIT Press

Stoneman P, Battisti G (2010) The diffusion of new technology. In: Bronwyn HH, Rosenberg N (eds) Handbook of the economics of innovation, vol 2. Elsevier, pp 733–760

Vetterli C, Brenner W, Uebernickel F, Petrie C (2013) From palaces to yurts: why requirements engineering needs design thinking. IEEE Internet Comput 17(2):91–94

RE-DT-UX: Moving from a Discipline-Based Approach to a Role-Based One

Kerstin Roese, Katharina M. Zeiner, and Rainer Wasgint

Introduction

Our research group at Siemens AG is comprised of 25 User Experience Designers, Design Thinkers, and Requirements Engineers. We believe our interdisciplinary teams allow us to better serve our customers at Siemens. This is not because this might allow us to have experts focus on their specialties but because working as T-shaped teams allows us to benefit from a multitude of experiences and opinions while developing a great user experience for our software solutions. Rather than focus on our original disciplines, however, we have found that thinking in roles that can be filled by different individuals when needed allows us to better respond to quickly changing project requirements.

In this chapter, we describe how the various experience-related roles work together in our team and how we use them to best support the development of products with a positive UX.

Joining Forces to Create Offerings with Great UX

Our goal is an efficient design for offerings that create a great User Experience (UX). UX is about the entire experience in connection with the product or service under consideration. In this respect, UX encompasses more than usability or UI, where

K. Roese (✉)
T RDA SSI UXD-DE, Siemens AG, Erlangen, Germany
e-mail: kerstin.roese@siemens.com

K. M. Zeiner · R. Wasgint
T RDA SSI UXD-DE, Siemens AG, Munich, Germany
e-mail: katharina.zeiner@siemens.com; rainer.wasgint@siemens.com

J. Hehn et al. (eds.), *Design Thinking for Software Engineering*, Progress in IS,
https://doi.org/10.1007/978-3-030-90594-1_6

only the direct interaction is considered. The term experience implies the integration of all relevant aspects, including the information architecture, the performance, the content presented, etc. It is therefore important for us to involve the customer and all stakeholders in the design process at an early stage. The Customer Experience can already be influenced with the idea and the definition of an eco-system (for more information on this approach see Lowdermilk T and Hammontree M (2020)). This is only truly possible if we approach the user's needs empathically from various directions.

Tim Brown who coined the term T-shaped people describes them as follows: "T-shaped people have both depth and breadth in their skills." Rather than being only experts in their field, "the horizontal stroke of the "T" is the disposition for collaboration across disciplines. It is composed of two things. First, empathy. It is important because it allows people to imagine the problem from another perspective-to stand in somebody else's shoes. Second, they tend to get very enthusiastic about other people's disciplines, to the point that they may actually start to practice them" (Hansen 2010). This is what our current role-based approach at Siemens T allows us to do.

Our basic understanding as UXD at Siemens Technology is that we work in the field of customer experience management. This means that the user experience must already be taken into account while devising the business model, with the definition of the eco-system, and all other phases of the product lifecycle.

Everything is part of the whole. So, all methods and tools used serve the overarching goal of finding out the needs of the user and to find "the best fitting way" to implement them. It is important to ensure that we are talking about analyzed real user needs and not user wishes. This is where Agile RE comes into play. In Agile RE methods and best practices of traditional RE are partially incorporated into the tasks and roles of agile development without violating agile principles. Here, traditional parts of the traditional RE approach are used to increase the description quality of Epics, User Stories or Feature definition. For example, wishes identified in interviews are often equated with needs, without trying to understand the intention of the user or trying to identify the potential and tasks behind the user's wish. This could be a user insisting that a button 'has to be red' and the design team blindly implementing a red button rather than asking why this is so important for the user. It is like staying at the symptom level in medicine without asking about the causes. In our button example, this could, for example, be caused by the user wanting to make sure that they always know where this specific button is located. This can be implemented with a red button but there are also other interaction and visual design choices that can fulfill the same need for safety or competence. A good user experience can arise only if we try to understand the needs behind these wishes (for more on this see Hassenzahl 2008).

Our "Old" Discipline-Based Approach

In the past, we saw ourselves as different disciplines working in one team. This included Usability Engineers, Design Thinkers, Requirements Engineers, and others.

Usability Engineer

A Usability Engineer drives the research strategy. By conducting usability studies and performing contextual inquiry he analyzes the user's workflow. This allows him to understand the user and customer needs. The Usability Engineers is also responsible for testing. He evaluates prototypes, runs User Acceptance Tests, and performs heuristic evaluations. He is also responsible for the interaction modeling. This includes considering mental models and task models, creation of storyboards and user journey maps. After initial testing of his concepts, he also creates wireframes that visualize the concept.

Design Thinker

The goal of a design thinker is to establish a human-centered mindset within the company. They plan and execute Design Thinking projects that develop innovative products, and services. A good Design Thinker creates an understanding of a customer or user, their pains and gains, as well as their needs. They do so through empathy building and use those insights to iteratively generate ideas for innovative concepts that are tailored to user's needs.

Requirements Engineer

In our understanding, a traditional Requirements Engineer's main task is to establish a common understanding between the stakeholders of the project. He not only bridges the gap between business and technology, but also reaches out for the agreement between all involved parties. In agile teams, he accompanies and supports the design sprints from Design Thinking, while, for example, applying and enforcing traditional RE best practices, like precise, testable statements and using the artifacts created in a structured way.

As more and more of our Siemens internal customers started adopting agile, some even moving to DevOps, we noticed that this discipline-focused approach was not serving us well. Therefore, we made the shift towards a role-based approach.

Why We Need a Unified Approach

When it comes to Agile SW-development, tasks and responsibilities associated with development are typically distributed over different development team members by roles. Besides developers, typically roles in the projects we work on include Product

Owner, Architect, Requirements Engineer, User Experience Designer and of course others. However, due to the ever shorter release cycles of SW-Products releases, we have observed and experienced the following pitfalls, both within and outside our organization, that work against our goal of creating great UX. In this section, we will describe these pitfalls and how they affect the overall development goal.

No or Very Limited Market(segment) (Over)view, No Business Model, No Bundling, Vision and Scope (V&S) Not Existent or Not Agreed, Stakeholders Not Fully Interviewed

Some products that are being developed are technological solutions but do not address a specific market/customer/user need and do not have a vision of what they are trying to address. The general development approach itself, be it Agile, Scrum, SAFe™ or DevOps is often used as an excuse for not applying certain development best practices. For example, not fully described and agreed on, realistic and believable project vision exists between stakeholders and team members. Therefore, there is also no clear scoping of such a vision with respect to a given market or customer. On the contrary, the market and customers segments have not been investigated thoroughly, described and neither is the resulting product agreed to be positioned well within the known market. This often means no business model is defined and described and its interaction with the intended product/customer is therefore unclear and the overall applicability has not been thought through. This would not be a problem if there was a single strong product visionary/product lead. In some projects, even if benchmarks and market research are performed, they are not shared with the project team in the project context with respect to the product's features and unique selling points. On the contrary, often due to department splits or responsibility splits also little to nothing is described with the respective domain expertise at hand. This means the positioning of the product with its intended features and bundling is completely unclear beforehand. All this often results also in no pilot or lead customer being included in the project, more or less leading into the next area of problems.

Non-existing Lead or Insufficient Customer Availability

This lack of research or overview in turn often leads to an unclear mission statement of the project's development goals and no real timelines that can be addressed by an Agile development methodology. This is complicated further if sprints are conducted delivering features of unclear customer value. Furthermore, in today's projects, some crucial aspects and preconditions necessary for a truly Agile development approach are often not met. For example, the customer is not at all or not really permanently on board, thus resulting in the development team being unable to resolve issues. This results in time delays or the team making decisions. Overall, this makes the work of the UX Researcher infinitely harder since it is that much harder to develop insights on development results Of course, one could argue that the Agile approach has a fail-safe for this at demos and that deliveries can still be rejected. Nevertheless, in larger organizations, this leads to huge inefficiencies and other roles taking over the roles of customer or even product owner often leading to sub-optimal decisions.

Missing Domain Experience and Domain Expertise

Often enough, the person executing the role of Product Owner is also unsure in details of the domain and thus not able to resolve such questions either. In some cases, product ownership is not associated with deep domain knowledge. Hence the product owner also needs to rely on the results from User Research which are hampered by the lack of access to users. Typically, this deficit in domain expertise leads also to the requests of Product Owners towards architects to decide on the product's functions and features via selection of technologies/components. This approach automatically limits the feasibility and applicability of the product's solution scope. Alternatively, the burden falls on Requirement Engineers to define the functionality with some creativity at hand. Especially in projects when UX is only included once development has already started this means it is often too late to make changes in features and functions that drastically affect the underlying architecture.

Missing Customer Insights: User Experience Not Investigated Due to Misprioritization, Too Little Knowledge on Importance of UX

Last but not least the lack of awareness for the insight of customers' real wishes and needs and their intended interaction ala Design Thinking or the elaboration and design of customer experience in an Agile manner is often neglected either due to time pressure, the team not being able to apply the methods, not knowing them at all or the misfit of such methods with existing roles and the development approach. This could be that no UX expert is part of the development team or the methods required do not fit into the sprints. This often means even if they are applied, there is not enough time to deliver the intended improvement in the product's interaction design and the resulting user experience.

Missing Strong Lead/Visionary

All of these described pitfalls probably would not derail a project if there was a strong lead or visionary, vouching with single responsibility for the product's vision, scope, and technology to be developed/applied, like Steve Job's iPhone idea, and willing to accept and carry the associated development risk, having entrepreneurial spirit and taking the necessary involved business risk from cradle to grave. Such a single strong product lead role typically does not exist in a large organization's development either. Rather, there is a split of associated responsibilities over several departments, roles, or boards as well as risk averseness, in combination counteracting real Agility. Not to mention, political aspects such as the blocking of knowledge between organizational units. Most often, we do, however, simply observe a lack of time for intense, continuous collaboration between different organizations/departments like marketing, sales, and development.

While we cannot single-handedly address all these pitfalls, we have noticed that making small changes in how we, as a team, work in agile projects can lessen their effect.

A Role-Based Approach

Our experience over the past two decades has been that agility has changed our work environment dramatically. In a way, software development's mindset has moved closer to our mindset. At the same time, strategies that allowed for an integration of Requirements Engineering, Design Thinking, and User Experience into the development process have become harder to implement, if not obsolete. We have found that shifting to a role-based approach has helped us adjust to this change in our environment as well as tackle the challenges we observe in agile SW development. Figure 1 shows the different roles we work with. Note, that not all roles are present in every project, and we also have times when not all roles are filled within our research group, but we believe that they all play their role in delivering great UX.

Figure 1 describes the different roles we use. Generally, they can be grouped based on their focus. The first is on the What ("What should be developed?"). These roles are especially active at earlier phases of product development. These roles are Experience Analyst, User Researcher, Business Analyst, Information Architect and Content Strategist. Somewhere in the middle, we have our UX Strategists. The other focus is that on How ("How should the thing that is being developed look and interact?"). These roles are more active during the elaboration and construction phases of a project. These roles include Interaction Designer, Accessibility, Visual Designer, Front-End-Developer, and UX Tester.

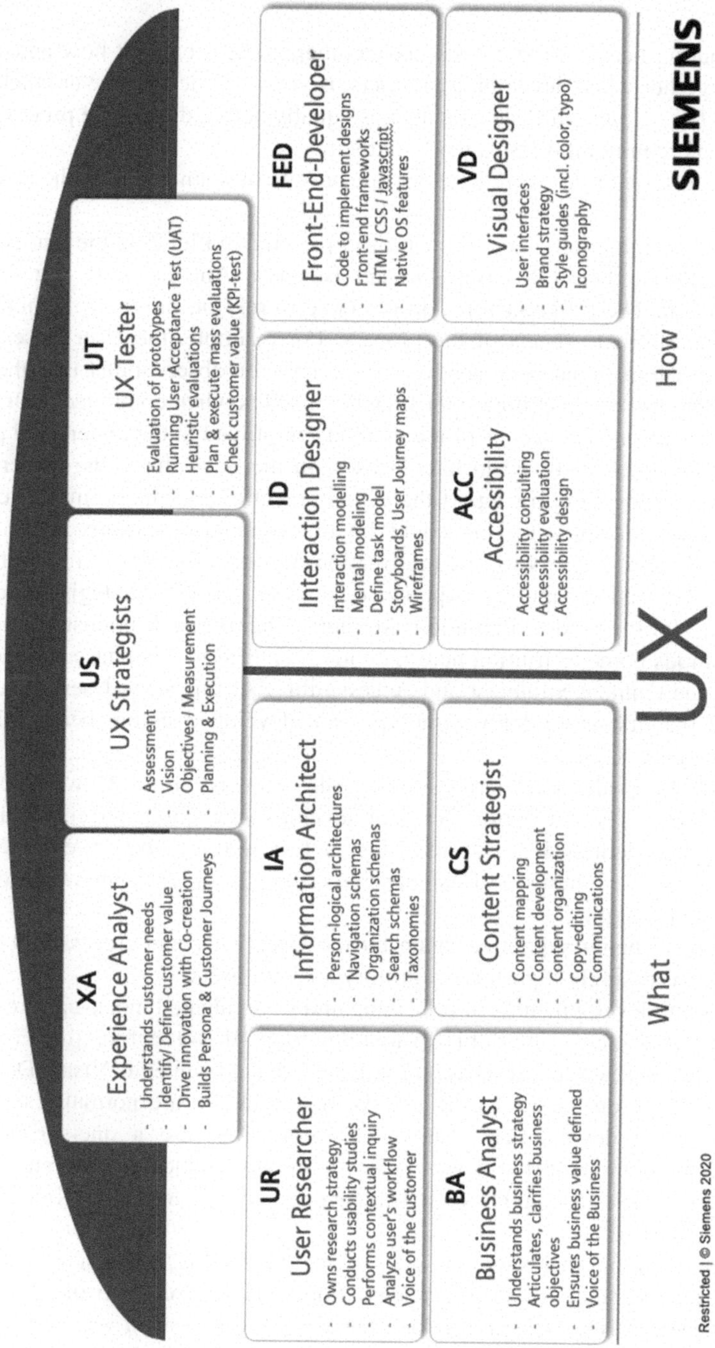

Fig. 1 This diagram shows the different specialties that work together to create great user experience

How They Work Together

By differentiating between these roles, we can then make it explicit how and when they work together when developing products. If you take the average architectural process (see Fig. 2), not all Roles are always equally active during the process and not all projects require all roles to work on them.

While the different roles work together one or two roles tend to be in the lead at a given time.

If you take the initiation phase of the project pictured in Fig. 2. In the Understand phase, User Researcher, UX Strategist and Business Analyst work together to develop a vision. The different roles do not have to rely on different methods but can focus on different aspects of the process. For example, the User Researcher collects information about user needs in interviews or observations that the UX Strategist might use to glean important strategic insights when combined with other pieces of information. The results of the methods employed in the understand phase are used by the roles involved in later phases. In the next phase, the Experience Analyst drives innovation to support the development of an opportunity statement that actually creates impact. This could be in co-creation sessions or in more "classical" DT workshops. Next, the Information Architect focuses on requirements that can then be used as input by Experience Analysts and UX Strategists to create concrete ideas for the product definition. Amongst other methods brainstorming and ideation methods come to fruition here. The ideas created in this phase can then be checked by the Content Strategist and Accessibility. At this stage User stories are defined and the first low-fidelity prototypes are developed before being initially evaluated by the UX Tester.

After this first evaluation loop, the other roles focused on the How to become much more active. The Interaction Designer develops interaction concepts while the Visual Designer defines UI patterns. This is followed by the development of wireframes from the Interaction Designer while the Visual Designer works on the UI concept and design.

Here they are being supported again by the UX Tester and Accessibility who evaluate the increasingly high-fidelity prototypes with users.

At this point, development is in full swing and the evaluated mock-ups are being implemented this means after this evaluation loop, if not before the Frontend Developer becomes active and starts building the frontend while the Visual Designer continues working on the visual design. At this point, we are normally sprinting together with the development teams and most often use a one-sprint-ahead approach (Alt-Simmons 2015) if we are not more systematically integrated either through Dual Track (Sy 2007) or a framework such as SAFe™ (Knaster and Leffingwell 2020).

There are two/four other roles that were not included in Fig. 1 which both become relevant if we are supporting agile release trains. Those roles are the Technical Product Owner (TPO) Design Product Owner (DPO) as well as their Project Management Office (e.g., PMO Design) counterparts. With these roles, the other

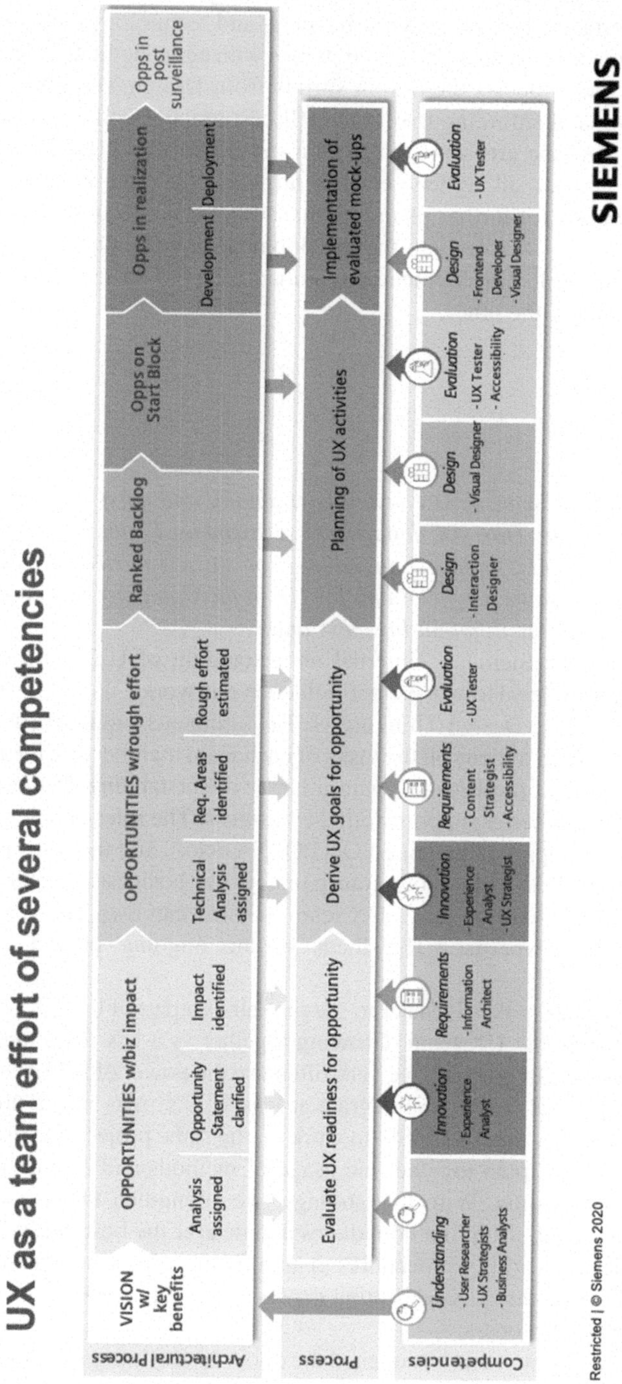

Fig. 2 The different roles work together throughout the development process. Restricted | © Siemens 2020

roles are directly embedded in the agile development process. The TPO, for example, not only interfaces the gap between business and technology, but also reaches out for the agreement between all involved parties with agile methods. Along the line he accompanies and supports the design sprints from Design Thinking, while, for example, applying and enforcing traditional RE best practices, like precise, testable statements and using the artifacts created in a structured way. The TPO keeps the overview and manages all agilely created artifacts, that are in later stages of development the basis for further planning, managing, and execution of the project.

This integration of UX roles into the development process allows us to leverage the benefits of both agile development and human-centered design without sacrificing one for the sake of the other.

Conclusion

The working world at Siemens is constantly changing and must also cover various customer requirements. Our UX department at Siemens Technology has been for more than 25 years, and a lot has changed here too since it started with Ergonomics evaluations. It has become a central hub for UX in the Company and is offering UX consulting services to the different business units.

If you look on our Siemens—internal understanding of UX—you see a huge number of different roles. How does that influence our work? We no longer discuss various disciplines like: Design Thinking vs. Requirements Engineering. We try to avoid these terms. With a focus on the tasks of each team member and the role he/she is playing in a specific project—we gain a better understanding of the interaction with the customer and better support from the business. The roles enable us to match better to standard IT approaches (like PO/TPO/DPO/etc.), and the roles give a clear frame of responsibilities and more transparency for business partners. Another relevant point is the situation, that a UX team member can own different roles—in dependence on his expertise. The focus of the ongoing project defines his primary role.

Talking about roles makes it easier to see oneself as a part of the UX team instead of being an expert in the UX team. Growing together as a UX team means also to learn from each other. In practice, we very often use the power of a 2-expertises-UX-consulting. That means we try to integrate at least 2 UX roles into a project. That gives the 2 UX Experts the possibility to work on the same project and exchange the different expert views, plan together the usage of methods and discuss about pros/cons and expected results. With this strategy, we strengthen the expertise of the whole UX team, avoid discipline boundaries and deliver the best possible result for the customer. These working procedures enable us to support the customer with a long-term and high quality UX consulting expertise along the product definition and product development process.

Over the last few years, this has opened the doors to very early integration into the product definition phase, we are now a partner for the development phase and work with and in development teams. Being a partner in the development process—and not only a temporarily expert—that was the target we were aiming to achieve with this approach.

Outlook

You might ask where we want to focus next. These days, it is more essential than ever before, to communicate, to build expert communities inside the company and to coordinate the own work with inner source models, internal exchange events and support of ongoing learning like a weekly best-practice exchange. We believe these kinds of activities allow us to increase the UX maturity of our entire corporation. Our UX department is realizing this in our UX CAMP. UX Camp is a creative space but also a virtual UX community hub. We offer weekly Best-Practice-Exchange, free "first-aid" support, regularly events for experts and management, and free usable method sources for all business departments inside Siemens. This affects a growing UX culture, growing UX maturity and over all the focus on customer needs and values.

References

Alt-Simmons R (2015) Agile by design: an implementation guide to analytic lifecycle management. SAS Institute Inc., Cary

Hansen MT (2010) IDEO CEO Tim Brown: T-shaped stars: the backbone of IDEO's collaborative culture. Chief executive. https://chiefexecutive.net/ideo-ceo-tim-brown-t-shaped-stars-the-backbone-of-ideoaes-collaborative-culture__trashed. Accessed 10 Feb 2021)

Hassenzahl M (2008) User experience (UX) towards an experiential perspective on product quality. In: Proceedings of the 20th conference on l'Interaction homme-machine

Knaster R, Leffingwell D (2020) SAFe 5.0 distilled: achieving business agility with the scaled agile framework. Addison-Wesley

Lowdermilk T, Hammontree M (2020) The customer-driven culture: a Microsoft story: six proven strategies to hack your culture and develop a learning-focused organization. O'Reilly Media, Newton

Sy D (2007) Adapting usability investigations for agile user-centered design. J Usability Stud 2(3): 112–132

Understanding the Introduction of Design Thinking as a Change Process

Martha Fritsch

Introduction

If the world we live in was previously complex and dynamic, it now becomes "VUCA" (volatile, uncertain, complex, and ambiguous) (Eggers and Hollmann 2018). This means companies today are forced to think innovatively to reach their customers. Developing innovative and customer-oriented products is the commandment of the hour and Design Thinking is an excellent way to ensure this. So why does not every company use Design Thinking in software engineering to identify requirements and then, at best, dip into agile software development? One reason is the implementation of Design Thinking in companies and their corresponding prevailing software development processes is not trivial. It is an intervention in an established process, leading to profound changes in various areas. Preconditions have to be established, processes have to be adapted and, in addition, a change in the mindset of those involved in the development process is needed.

When Design Thinking is introduced into software development projects, this often happens in an unstructured way and aims were formulated vaguely. It is usually unclear where changes should be made within the software development process. Often the introduction is not initiated by management, but by an employee and tested on a project in an unsystematic way. Usually, such test projects fail and the introduction of Design Thinking deemed failed.

However, in this contribution we will not discuss how Design Thinking is integrated into software development, but rather address the integration process itself. The present text is intended as a contribution to comprehend the introduction of Design Thinking into software development as a change process, in order to be able to meet the challenges that arise during implementation in a more targeted manner and to provide a structure for the integration of Design Thinking into

M. Fritsch (✉)
Hemsbach, Germany

J. Hehn et al. (eds.), *Design Thinking for Software Engineering*, Progress in IS,
https://doi.org/10.1007/978-3-030-90594-1_7

software development. To shed some light on these issues, the contribution begins with a case study. The following chapter will explain the meaning of Change Management. Furthermore, two models are explained, which illuminate the change process from different angles. Afterwards, the topics of resistance and communication connected to change management are examined and finally a conclusion is drawn.

Case Study

Company alpha is a medium-sized service company which, besides other things, programs software on demand. In addition to classic software development according to the waterfall model, some development teams work according to SCRUM. This was seen as a first step to increase the company's degree of innovation and to bring more speed and flexibility into the development process. The main challenge of *company alpha* is that neither the development time nor the quality of the resulting software products can compete with those of small, flexible, and innovative software companies.

As part of a company-wide change process, many new ways of working are tried out and there is basically an open climate that encourages individual employees to go new ways and try out new methods. In this context, one of the team leaders had the idea of integrating Design Thinking into agile software development in order to keep the products competitive and to focus more on customer benefits. So, a small group of design thinkers emerges that feels capable of holding workshops and promoting the topic. The idea of starting a Design Thinking Challenge for an outdated product is approved by middle management, who agrees to release the participants from their actual tasks for the duration of the two-day "test project" workshop. Highly motivated, special rooms are rented, the participants are selected and invited, and the workshop is held. The atmosphere during the workshop is good, and the participants can easily get involved in the challenge and their tasks. The results of the Design Thinking Workshop are excellent. A number of new, innovative requirements for the aging product emerge, which have to be checked for feasibility and economic viability in a follow-up step and communicated to the teams.

With this success behind them, Design Thinking is used in parallel for other products with other challenges. Unfortunately, the workshops cannot follow up on the success of the first workshop "Test Project." New participants were invited for the new challenges, but they already heard rumors about the first workshop. The atmosphere was far from good, and the first participants started to check their e-mails during the workshops or to extend the breaks significantly. It turns out that some participants attend the workshop just out of curiosity to see if anything changes in requirements engineering. Other voices are raised that the participation is only based on getting 2 days of diversion from the daily work routine. Still others participate only because they were sent by their supervisor. In many cases, the impression arises that the participants are not at all aware of why they are taking part at all, what the purpose of the event is, or what consequences arise from Design Thinking for the individual.

Within the management, voices are getting louder that question the sense of Design Thinking. No more workshops are held. The results of the past workshops are swept under the carpet and Design Thinking is no longer followed as a method. The reasons given for the failure are that Design Thinking simply does not fit into *company alpha*, that it is too time-consuming and expensive, and that the results are too creative.

What happened between the two workshops? Why did the general mood change? The rumor spreads among the staff that Design Thinking would disrupt the development process and that work might have to be done more than once. It was also speculated that the requirements that arose could not be implemented and were therefore rather visionary. Not tangible, not concrete enough. Within the management, there was a fear that the cost framework would be blown up by the multiple iterations and that the results would not bring the expected ROI.

In this case, the introduction of Design Thinking was attempted bottom up. Although the introduction was certainly motivated and also well-intentioned by the core team, besides the weak technical introduction, it is mainly the lack of change management that causes the introduction of Design Thinking to fail.

What Does Change Management Mean?

To comprehend the introduction of Design Thinking into software development as a change process, it must first be clarified what change management exactly means. Basically, it is not about defining the content of the goal, but rather about designing the path from the previous way of collecting requirements to the new way of collecting requirements by using Design Thinking and feeding these new requirements into the development process (Lauer 2019).

The implementation of Design Thinking in, e.g., *Scrum* is a change in process organization. A software development process is changed, and measures are developed to create a new process. However, this point of view only refers to the technical side of the change. Its interdisciplinary aspects are represented by the soft processes, triggered in those involved, namely management and development team. For a successful introduction of Design Thinking, it must not only be implemented in terms of processes, i.e., in a technically adequate way, but also in a cross-disciplinary way (Stolzenberg and Heberle 2013). Above all, the interdisciplinary aspects within a change project decide on success or failure. Interdisciplinary aspects include whether the new approach is accepted, whether its necessity is accepted, and whether the willingness to actively participate in the Design Thinking process by the team is given (ibid). Especially in well-rehearsed teams, resistance can arise on the interdisciplinary side of the change. In the following, when speaking of change management or change projects, it will, therefore, refer to the interdisciplinary part of the change.

3-Phase-Model According to Kurt Lewin

In the 1940s, Lewin (1963) already describes a 3-phase-model of organizational development which can be applied to all change processes (Lewin 1963). He assumes that there are two forces in the framework of change processes: on the one hand, there are *restraining forces* that represent the fear of change, i.e. in the present case the fear that Design Thinking will change the tried and tested form of raising requirements. On the other hand, there are *driving forces* making a change necessary, i.e. market pressure, pressure to innovate, or, as described above, altered user requirements for the product. Lewin assumes that an imbalance between the two forces must be created with the purpose of bringing about a change. Resistance must be kept as low as possible in order to secure a successful change.

Lewin divides the change process into three parts. In the first phase (*unfreezing*), a previously established initial state is thawed. In this phase, the change is usually announced, and the software development team learns about the introduction of the new way of requirements engineering. In the case of *company alpha*, mentioned in the case study, this phase was underestimated. The communication of the possible change was not initiated by the management, but at the employee level. Accordingly, Design Thinking was given little weight from the very beginning, even though it was explicitly welcomed by the management. In the second phase (*moving*) the desired state is implemented, meaning that the technical connection between Design Thinking and the software development process is set up. Solutions are found to integrate Design Thinking into software development (Baltes and Freyth 2017). As can be seen in Fig. 1, a drop in performance is expected in this phase. Having said this, it is

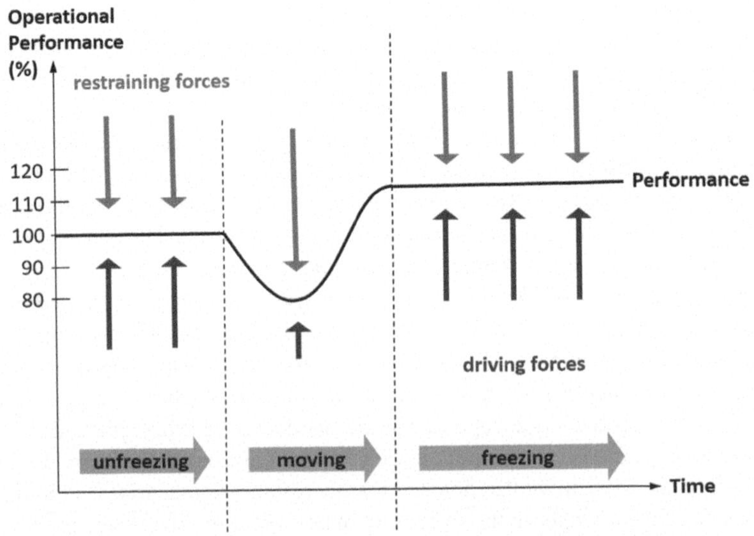

Fig. 1 3-Phase-model according to Lewin (own presentation following Lauer (2019))

advisable to test and practice Design Thinking. It makes sense to carry out test projects of less economic impact. In the case of *company alpha*, in fact a test project with a low economic impact was chosen. But due to the missing management communication, the signal was the wrong one. The impression was created that with the unimportance of the product, the new way of requirements engineering was also unimportant. As a result, the last of Lewin's three phases never took place.

In this third and last phase (*freezing*) the new way of gathering requirements is consolidated and works—at best—on a higher level as before (ibid).

What is so simply described contains a very important idea for the introduction of a new method to gather requirements. A change process has a beginning and an end. Changes have to be prepared and after being made they have to be consolidated (Kaiser and Schwertner 2020). This is the only way to create new routines among the team. When change processes seem interminable, the persons involved quickly become overtaxed and resistance develops which could be avoided at this point. During the change process, the individuals are expected to do a lot. Especially very structured employees, as they are prevalent within IT, often find it difficult to step out of their comfort zone. A clear start and end point create certainty that the change occurs within a manageable time frame. This also motivates those involved to contribute a certain amount of additional work (ibid).

7-Phase-Model According to Streich

As mentioned at the beginning of this contribution, the interdisciplinary part of the change process has a special significance. This primarily means to devote oneself to the people involved in the change process. The successful introduction of Design Thinking into software development depends largely on the support of the people who are directly affected by the change. Every change, especially abrupt ones, leads to reactions on the part of those affected. The 7-phase-model according to Streich (2016) describes the emotional reactions of those involved in the change. Depending on the temporal sequence, the perceived competence of each individual is altered during the change process (Kreutzer 2018). The management task is to recognize where individuals are on this way and to guide them to the next developmental stage within the change process (Baltes and Freyth 2017).

Usually, every change passes the following seven phases: Shock, Negation, Rational understanding, Acceptance, Exercising, Realization, and Integration. The newer the change is, the stronger the resistance (negation) against the process itself and the behavior to be modified will be. For example, the more an organizational culture suppresses fault tolerance, the more the exercising phase is torpedoed (Streich 2016). The seven phases described by Streich (2016) are explained in the following and their interaction is illustrated in Fig. 2:

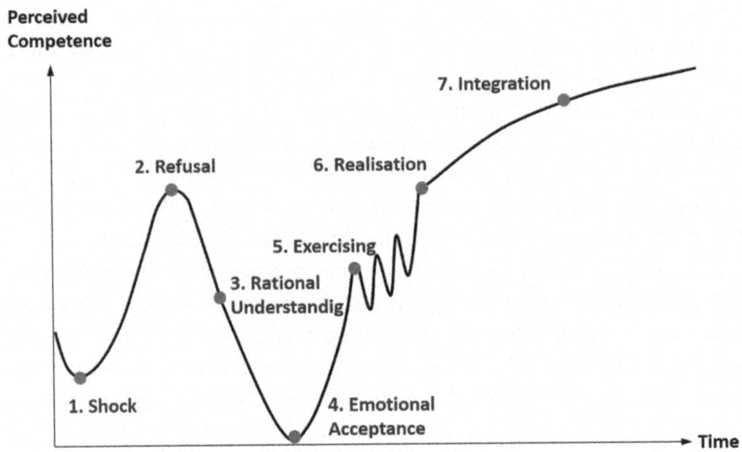

Fig. 2 Change curve (own presentation following Kreutzer (2018))

1. *Shock:* Especially in innovative change projects, an initial phase of shock characterizes the emotional reaction of those involved. Depending on the extent or characteristics of the shock, it can be smaller or larger.
2. *Refusal and negation:* The next phase is marked by fears, annoyance, and active rejection of the change. For less active people, this may simply be a passive attitude or frustration.
3. *Rational understanding:* At this point, most of those involved can already understand the reasons for the change. This means that the change is tolerated by now, but yet there is no deeper willingness for a personal change.
4. *Emotional acceptance:* The change is now accepted and approved. The people involved are aware that they themselves must change something. In this phase, fears may arise that the new task or role cannot be fulfilled well.
5. *Exercising:* The employees start to actively deal with the new requirements. They begin to change their behavior, experience successes and setbacks. If the setbacks or negative experiences are too great, there is a danger of falling back into phase two.
6. *Realization:* In this phase the realization grows that the changes have indeed positive aspects. The resulting positive basic attitude promotes further changes in the behavior of those involved and the change becomes more and more integrated.
7. *Integration:* Change has become the norm. The new routines are carried out as a matter of course and integrated into everyday life.

Against the background of the phases of change shown in Fig. 2, it is substantial to schedule enough time for the integration of Design Thinking into software development; not only to test the method, but also to grant the participants enough time to process the change emotionally. Only in this way, the new approach can be consolidated and at some point, be considered as a normal way of requirement incorporation.

If we look at the model, we see that the employees of *company alpha* could not leave phase two "refusal." The negative attitude within the second workshop shows this quite clearly. This negative attitude then spread beyond the second workshop to others and influenced the entire introduction. For this reason, the topic of resistance will be examined in more detail here.

Resistance

A critical phase is the second phase within Streich's change curve, which is about negation and resistance. According to a study by Capgemini Consulting (2012) "realizing and meeting resistance" is the fourth important success factor to accomplish a change (Capgemini Consulting 2012). Particularly against the background that resistance is not always directly recognizable as such, special attention must be paid here. First of all, it must be stated that resistance per se is not bad. It is rather a natural reaction to changing conditions (Dahms 2010). Quite the contrary, it should be regarded critically if resistance is missing, since that might be a sign that the group does not believe in the introduction of Design Thinking at all.

In order to meet resistance, the organization should be aware of the different forms of resistance. According to Doppler and Lauterburg (2008), resistance can be divided into four different manifestations (Doppler and Lauterburg 2008). As shown in Fig. 3, the characteristics range from active counter-arguments to in-attention, fatigue, or illness. The last mentioned are often difficult to identify as an expression of resistance. It is, therefore, all the more important to be attentive to be able to meet even quiet resistance. Quiet resistance seems uncritical at first glance. However, it is more difficult to deal with it. To meet open protest is easier, because it is obviously recognizable as such. Resistance often contains a hidden message and is not related to change. The reasons for this are to be found in the emotional, interdisciplinary area.

To look closely is important, because Design Thinking thrives on the people who work with it. If these people refuse the implementation, it is hardly possible to introduce it successfully. In many cases, ignoring resistance leads to blockages and thus to failure, as in the example of *company alpha*.

Resistance can be subdivided according to whether the motives of those affected are rather objective or based on power interests. Professional resistance is more likely to be found at employee level. In concrete terms, concerns arising from a professional point of view are mainly based on fear of being overwhelmed, criticism of the working methods, or even the loss of a job due to the change. Resistance based on power interests is more likely to be found in middle and top management. Fears arising from power interests are particularly directed at the loss of influence or reputation or the loss of resources such as personnel (Lauer 2019).

	verbal	nonverbal
active	**Ojection**	**Excitement**
	Counter-argumentation Accusations Threats Polemics Stubborn formalism	Unrest Dispute Intrigues Rumors Clique formation
passive	**Evasion**	**Listlessness**
	Silence Trivialize Fool around Ridicule Unimportant debating	Inattention Tiredness Absenteeism Inner emigration Disease

Fig. 3 Resistance (own presentation following Doppler and Lauterburg (2008))

In the case of *company alpha*, the resistance of the workshop participants is shown subliminally by the extension of the breaks or the checking of e-mails. At this point, it would have already been possible to intervene and seek the dialogue with the persons concerned.

If resistance is ascertained, it should always be reacted to and intervened. However, the timing of intervention plays an important role when it comes to meeting resistance. According to Dahms (2010) two main mistakes are made when dealing with resistance in groups:

- The leadership reacts too quickly to the resistance although the group would still be able to work. This can disrupt group dynamics and the resistance to Design Thinking increases.
- The leadership reacts too slowly, although the group has already lost its ability to work. Due to the lack of intervention, there is a risk that the ability to work will decrease further (Dahms 2010).

Resistance can be based on the fear of not being up to the new situation. Employee training helps to build up necessary skills and simultaneously reduces resistance. Knowing the background and theory of Design Thinking can help to reduce resistance to its implementation. Thus, building competence not only promotes development and gaining skills but also has a function in resistance management.

Resistance should be channeled, no matter if it arises within the group or if it is based on power interests. It is absolutely necessary to give space to the employees to express themselves and to reduce pressure. Communication is the instrument of choice to involve employees in the change process, thus increasing the probability that they will support the introduction of Design Thinking.

Communicate Change

Communication should not only be considered when resistance toward the planned introduction of Design Thinking in software development arises. Change projects question the current conditions and rearrange them. Accordingly, this creates uncertainty among employees and questions arise (Stolzenberg and Heberle 2013). What will the new process look like? Which colleagues are involved? What consequences will this have for my everyday life? To answer these questions in a structured way, it is often worthwhile to design a plan of communication measures along the introduction, which is created initially and continuously adapted to new circumstances (Chiess 2016). A communication plan (see Fig. 4), including the communication needs, the time, the medium used, the aim and affected stakeholders, is a helpful tool in change management.

Looking at communication according to the phases of change, the initial aim is to clarify the vision and the background for the introduction of Design Thinking. In addition, the effects on those affected, expected changes and difficulties have to be described. During the implementation it is important to maintain the motivation of those involved. Primarily, it is about communicating successes, i.e., to spread factual information. But it is also about meeting resistance (Lauer 2019).

In the case of *company alpha*, there was already insufficient communication at the beginning of the introduction. The people behind the introduction were not able to explain Design Thinking in a way that it would be understood by the management. The importance of the opportunity of Design Thinking was thus possibly misunderstood. This could be due to the fact that the management failed to communicate the introduction widely. This would have given the workshops more weight and reduced the character of a test balloon. As a result of the lack of communication, the fears of the individual participants could not be removed.

Target groups	Communication needs		Aims of communication	Media	Times
	Concerns	Questions			

Fig. 4 Communication plan (own presentation following Stolzenberg and Heberle (2013))

Change communication must not be seen as a one-way street (Kaiser and Schwertner 2020). It is not about pure information, but about getting in contact with those affected and seeking dialog. The appraisal interview can be used here to provide a forum for the employee's need for discussion and to specify expectations (Dahms 2010). Usually, appraisal interviews are held once or twice a year to discuss work and tasks or cooperation. But especially in the context of a change process, it is worthwhile to conduct structured appraisal interviews to nip resistance in the bud. These conversations could have been used in *company alpha* in several places and at several hierarchical levels. Especially in the context that it is not a one-way communication. Fears could have been addressed and resistance could have been overcome. Also, after the discussion among the staff escalated and the rumor mill about the new confusion in the development process started, more conversations should have been held. Communication after the failure of the introduction basically did not take place at all. Successful introduction of Design Thinking requires openness and honesty, which can both be conveyed best through the spoken word (Deutinger 2017). Of course there is a potpourri of usable media, but, however, nothing is so enduring and explains better than personal communication.

Lauer (2019) sums up the importance of communication and sees it as a kind of catalyst for change management. Communication creates transparency, weakens resistance, creates motivation, and promotes social integration (Lauer 2019).

Conclusion

To comprehend the introduction of Design Thinking into software development as a change project can contribute significantly to success. This approach shows the necessary seriousness in dealing with the process change and signals this to the participants. The three phases of the change process according to Lewin show that performance varies within the change process, which is perfectly normal. The seven phases according to Streich (2016) shed light on the emotional development of the affected people the change managers have to deal with.

Change takes time. Especially when Design Thinking has failed in a test project, one should not give up immediately but search for reasons. Hidden resistance within the team might lead to failure. Employees need time to internalize changes. Considering the needs of people in change processes is very important. Resistance can be countered by clear, transparent, and above all understandable communication along the process.

The reason for a failure of introduction can be a mistake within the new process. And this is exactly where the big advantage of a planned introduction lies—the procedure is transparent and can be both, checked and improved. A review and improvement work if there is a grown error culture in the company, allowing to test and, if necessary, adjust.

It can also be helpful to engage an external coach. When you think of a Design Thinking Coach, the first thing coming to mind is a Team Coach accompanying the design team. Here, however, a coach is someone who accompanies the change process itself, from the mediation of the various principles, ways of thinking and working, and methods to the introduction process as such (Tschepe 2017).

It is important not to underestimate the introduction of Design Thinking. In many cases, the introduction of Design Thinking fails without finding the actual cause. This is often a consequence of the false assumption that the process change is too small to be granted sufficient attention. As a result, the reason for failure is sought within the Design Thinking method itself and is attributed to the flawed nature of the method. However, this is often a misinterpretation.

Of course, the failure of the introduction in the case study cannot be attributed exclusively to a failure of change management, but if more attention had been paid to the process itself and the interdisciplinary aspects, the introduction would have been much more promising.

References

Baltes G, Freyth A (2017) Die radikal neuen Anforderungen unserer Zeit und die Konsequenzen für Veränderungsarbeit. In: Baltes G, Freyth A (eds) Veränderungsintelligenz. Agiler, innovativer, unternehmerischer den Wandel unserer Zeit meistern. Springer Fachmedien Wiesbaden, Wiesbaden, pp 1–77

Capgemini Consulting (2012) Digitale revolution. Ist Change Management mutig genug für die Zukunft? Available at: https://www.capgemini.com/consulting-de/wp-content/uploads/sites/32/2017/08/change_management_studie_2012_0.pdf. Accessed 8 Jul 2021

Chiess S (2016) Change Management bei der Einführung neuer IT- Technologien. Mitarbeiter ins Boot holen – mit angewandter Psychologie. Springer Fachmedien Wiesbaden, Wiesbaden

Dahms M (2010) Motivieren, Delegieren, Kritisieren. Erfolgsfaktoren der Führungskraft, 2nd edn. Springer Fachmedien Wiesbaden, Wiesbaden

Deutinger G (2017) Kommunikation im Change. Erfolgreich kommunizieren in Veränderungsprozessen, 2nd edn. Springer Fachmedien Wiesbaden, Wiesbaden

Doppler K, Lauterburg C (2008) Change Management. Den Unternehmenswandel gestalten, 12th edn. Campus, Frankfurt

Eggers B, Hollmann S (2018) Digital Leadership – Anforderungen, Aufgaben und Skills von Führungskräften in der "Arbeitswelt 4.0" In: Keuper F et al (eds) Disruption und transformation management digital leadership – Digitales mindset – Digitale Strategie. Springer Fachmedien Wiesbaden, Wiesbaden

Kaiser M, Schwertner N (2020) Change Management in der Kommunikationsbranche. Veränderungsprozesse in Medienunternehmen und in der Unternehmenskommunikation. Springer Fachmedien Wiesbaden, Wiesbaden

Kreutzer R (2018) Führungs- und Organisationskonzepte im digitalen Zeitalter kompakt. Agilität erreichen, Prozesse beschleunigen, Change-Management implementieren. Springer Fachmedien Wiesbaden, Wiesbaden

Lauer T (2019) Change Management. Grundlagen und Erfolgsfaktoren, 3rd edn. Springer Fachmedien Wiesbaden, Wiesbaden

Lewin K (1963) Geplante Veränderungen als Dreischritt: "Auflockern, Hinüberleiten und Verfestigen eines Gruppenstandards": Gleichgewichte und Veränderungen in der Gruppendynamik. In: Cartwright D (ed) Feldtheorie in den Sozialwissenschaften. Ausgewählte theoretische Schriften. Hans Huber, Bern

Stolzenberg K, Heberle K (2013) Change management. Springer, Berlin

Streich R (2016) Fit for leadership. Führungserfolg durch Führungspersönlichkeit, 2nd edn. Springer Fachmedien Wiesbaden, Wiesbaden

Tschepe S (2017) Was sind die wichtigsten Eigenschaften und Fähigkeiten von Design Thinking-Coaches? Erwachsenenpädagogischer Report. Humboldt-Universität zu Berlin

From Project Plans and Backlogs to Strategic Roadmaps: The Evolution Toward Value-Oriented Thinking in Requirements Engineering

Markus Guentert, Holger Rhinow, and Christoph Meinel

Design Thinking Is Desirability-Heavy While Requirements Engineering Is Feasibility-Heavy

The design agency IDEO popularized the notion of *innovation* as an intersection of three different qualities that need to be achieved: (1) desirability from a user's point of view, (2) viability from a business point of view, and (3) feasibility from a technological point of view. Agencies and education organizations such as IDEO and the Hasso Plattner Institute argue that Design Thinking addresses all three perspectives and therefore leads to innovative outcomes (Kelley and Kelley 2013).

Following this thought, the most promising ideas are the ones that solve a real customer problem, generate business value, and are technologically (and legally) manageable by the company. Especially agile approaches such as Design Thinking, lean startup, and Scrum propagate the idea of multidisciplinary teams that unite different skills to cover all three perspectives. However, this notion does not make explicit how these perspectives come into play during different stages of the development process. Are all three perspectives relevant in all phases or do they alternate? Can one perspective be conclusively defined at some point or do all perspectives represent recurring elements during the entire development process?

In our research on Design Thinking in practice (Schmiedgen et al. 2015) and in our collaboration with industry partners, we have observed that Design Thinking is mainly applied as a structural approach at the *fuzzy front end* of the innovation process. With Design Thinking, dedicated teams aim to explore problems of potential customers in situations that are of strategic relevance for the respective company. Design Thinking can therefore be understood as an approach to bring "outside"

M. Guentert · H. Rhinow · C. Meinel (✉)
Hasso-Plattner-Institut für Digital Engineering gGmbH, Potsdam, Germany
e-mail: christoph.meinel@hpi.de

J. Hehn et al. (eds.), *Design Thinking for Software Engineering*, Progress in IS,
https://doi.org/10.1007/978-3-030-90594-1_8

perspectives into the company (Brown 2008) in contrast to developing ideas only within the company itself. We refer to this as an "*outside-in*" approach.

Potential customers are hence interviewed or observed and studied in their underlying needs. Design Thinking encourages framing interesting observations as "problems" potentially worth solving. This process of explicating real-world problems constitutes a common starting point to find solutions. Based on the gathered insights, teams would then iteractively develop ideas, turn them into prototypes, and test their level of desirability with respect to the problems and needs of the targeted customers. After a few iterations, a team ideally arrives at a fitting combination of problem and solution that turns out to be desirable for the customer. Teams often refer to this milestone as "*problem-solution-fit*." The fit represents the team's knowledge, embodied in prototypes, and outlines a vision for the solution. Although this vision generally defines a strategic direction, it usually does not at this point describe a tangible solution design (which is concrete enough to be implemented). The creation of one would then follow in a subsequent stage of the development process, e.g., as part of requirements engineering.

Since the problem-solution-fit is a necessary interim result, we can argue that the desirability perspective is dominant during the Design Thinking stage. This seems reasonable as every company should aim to create and deliver solutions that address valid problems. Although this sounds very plausible in theory, many companies still fall through the cracks in this regard. It is reported that 7 out of 10 new products fail to deliver on their expectations, especially due to a lack of desirability (Green 2014). In conclusion, desirability is necessary—albeit not sufficient—for developing innovative products or services, so investing most of time and energy toward the desirability perspective in the beginning is valuable. We hence consider it good practice if a Design Thinking incarnation is desirability-heavy.

Whereas the outside-in nature of a Design Thinking stage pushes teams to regularly engage with customers, the subsequent *requirements engineering* stage does not necessarily foster this sort of engagement anymore. Customer interaction in the form of interviews or user tests may still happen "naturally" if this is a core belief of the company and also institutionalized well. For many companies, however, this may not yet be the case. In our collaboration with industry partners, we have observed that many initiatives that start with a commendable Design Thinking stage then turn into "submarine" implementation projects thereafter. This metaphor is derived from the idea that the company does not seek any sort of feedback from the outside anymore and is "diving under water"—with the risk of going in a subpar direction.

The nature of an initiative then shifts from an outside-in desirability-heavy to an *inside-out* feasibility-heavy approach as it enters the requirements engineering stage (see Fig. 1). Inside-out describes the dynamics in discussions that take place inside the company to determine the resulting product or service—instead of further focusing on a real-world problem of the customer. These discussions may evolve around feasibility questions such as compliance, legal implications, and security concerns or simply around subjective internal stakeholders' opinions. The same idea that was regarded as an opportunity for success from a desirability perspective, may

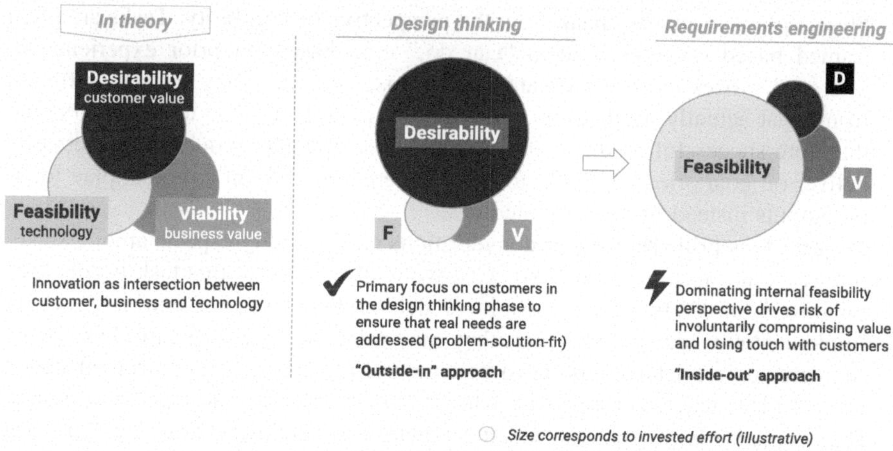

Fig. 1 Value perspectives on Design Thinking and requirements engineering

now be regarded as a risk of failure from a feasibility perspective. The feasibility perspective is certainly necessary to ground new ideas in the direction of a reality check. However, if a company operates in a one-sided feasibility-heavy manner, it may end up with undesirable outcomes. If, for example, a compromise is made in order to minimize development risks, this may result in a decreasing desirability of the solution. A critical decision for a compromise will not be recognized as such if customers are no longer involved in the development process.

While we endorse the desirability-heaviness of Design Thinking, we would like to point out the risk of a pontential one-sided feasibility heaviness in requirements engineering and subsequent stages of the development process. In this contribution, we aim to explain how the phenomenon of feasibility heaviness is caused by typical communication artifacts most often encountered in the industry, namely project plans and backlogs, and how this issue can be circumvented through strategic roadmaps. A well-designed strategic roadmap facilitates an outside-in perspective throughout the entire development process.

Project Plans Resist Adaptation to New Insights and Trigger the Wrong Conversations

Project plans typically come in the form of Gantt charts (Sutherland 2014). They layout planned tasks against an upcoming timeline, each activity being defined by its anticipated start and end date. Such project plans are intuitively understandable and are an often-encountered corporate practice. However, several concerns remain whether project plans are a sufficient means to deal with the complexity of implementing innovative products or services.

1. Project plans primarily speak from the perspective of feasibility. Tasks are often framed based on what a team "can do," e.g., based on prior experience or reference processes in the organization. This, however, may be very different from what actually needs to be done in order to deliver value to customers (and therefore successful products or services). Task orientation does not necessarily deliver on goal orientation. Or, to put it differently, tasks in a project plan focus on outputs instead of desired outcomes.
2. Project plans primarily trigger conversations on feasibility aspects among stakeholders who raise questions like "Can we really complete this task in this time window?" or "Who is responsible to deliver this task?" These are relevant implementation details; however, they may miss the bigger picture. Discussing implementation details appears to be a natural reaction when confronted with a project plan, whereas discussing whether these tasks are the right ones is not. Project plans often guide conversations on how to get things done (the "how") at the cost of understanding what needs to be done (the "what"). The success of the result in the end is, however, determined by the former rather than the latter.
3. Project plans do not invite a change of scope. A continuous adaptation to new insights is considered healthy in nearly every innovation process. It does not make sense to continue executing a plan which is already obsolete. Project plans, however, in their very nature suggest that the scope is fixed. Once a project plan has been communicated, it defines expectations on outputs for everyone involved. Even worse, project responsible people, e.g., project managers, are usually incentivized on delivering the project exactly as originally planned. Therefore, they are intrinsically not welcoming of change, even going so far as to resist it.
4. Finally, each innovative product or service should be designed around an overall life cycle (Moore 2014) rather than as a traditional project-like one-time effort with a clearly defined end date. Although the "main" development is considered done as soon as the product or service is initially shipped, the product or service itself should not be considered finished yet. On the contrary, after being introduced to the market, it needs to be refined according to market feedback and further enhanced to deliver new value to customers. Key resources, i.e., developers, designers, and product managers, are hence still required after the initial release and should not be regarded as freed up for new projects.

Backlogs Foster Reactive Bottom-Up Thinking Instead of Top-Down Value Orientation

A backlog can be considered an evolution to a traditional project plan which encourages change by design. The concept of a backlog became a prominent tool in agile software development processes such as Scrum (Sutherland 2014). In a typical backlog, backlog items—each usually corresponds to a product feature or product quality—are prioritized against each other in a dynamic list. This list is meant to be constantly re-prioritized based on new insights which are collected during the development process (e.g., as a result from user research or market feedback). Dedicated rituals, i.e., backlog refinements, trigger such a re-prioritization explicitly. Therefore, the scope which is managed by a backlog is not fixed. A Scrum team, for example, is by definition asked to regularly "check in" to see if what they are doing "is heading in the right direction, and if it's actually what people want" (Sutherland 2014, p. 9).

Furthermore, framing work packages as product features or product qualities supports thinking in results instead of tasks to be executed. This is in line with keeping an eye on the "what" instead of getting lost in the "how" which is a common critic towards project plans. If backlog items are additionally framed as user stories (which is optional, but an often-to-be-found combination), the desirability perspective is manifested in the process as well. User stories describe a requirement from the perspective of what a user wants to achieve (the "what") and the value it creates for them (the "why"). Whereas we argued that project plans mostly focus on feasibility, we can hence argue that backlogs focus on desirability. Even though backlogs mark an important improvement compared to project plans, they still come with four shortcomings:

1. Although backlogs are designed for prioritization, the granularity of individual backlog items tends to be very small so that they can be implemented within the capacity constraints of a single sprint (usually 2–3 weeks). However, if the product or service to be implemented is of a rather high complexity, this will result in numerous backlog items to be prioritized against one another. This circumstance, in turn, does not practically enable prioritization on a strategic value-oriented level, since a large number of items become cognitively unmanageable in their entirety. Speaking metaphorically, it is difficult to see the forest when standing in front of too many trees. Critical prioritization decisions should take place at a higher level of granularity, which many backlogs fail to reveal.
2. We often observe that backlogs foster reactive behavior when confronted with feedback. Once a new insight is gathered (e.g., through stakeholder feedback or user research), a corresponding backlog item is automatically pushed to the top of the backlog. The newness of an insight then subliminally triggers a perceived urgency for implementation, although this might be a subpar prioritization decision with regards to the whole. This "anti-pattern" requires an experienced product owner (or any other role that is responsible for prioritizing the backlog) and a holistic overview about the rest of the backlog in order to be circumvented, which may not always be the case.
3. Backlogs are steered from sprint to sprint (Sutherland 2014) and therefore encourage short-term rather than mid-term thinking. Backlogs let a team "drive by sight" instead of proactively driving strategic value goals. Whereas the next 2–5 sprints are usually planned to a certain extent, everything that follows thereafter remains nearly unknown or unmanaged. Requirements are collected "along the way." This mode of operations may be regarded as a bottom-up approach with only a vague overall direction. Again, prioritization is somewhat

dysfunctional if prioritization decisions only happen on the level of immediacy, but not in view of the whole strategy.

4. The bottom-up perspective of a backlog may result in a subpar phenomenon we refer to as the "watering can effect." Ideally, a product or service is grown along dedicated strategic goals, e.g., in the form of individual user themes which are rolled out consecutively. Each user theme represents a cohesive set of features that delivers recognizable additional customer value toward an overarching customer goal. When the customer uses the product or service again, they shall notice a "wow, this is new" effect. In this way, the marketing around the product or service can tell stories (e.g., through blog posts or email campaigns) and target new customers or re-engage with existing customers. The watering can affect, however, describes the opposite. Instead of focusing on one plant that is significantly grown with a large amount of water, all the plants in the garden are given water but each receives only very little. The result looks marginally changed and the product or service then tells fewer stories. Over a longer period of time, changes would still become visible, however, this approach unnecessarily constrains the marketing of the product or service.

Strategic Roadmaps Foster Value-Oriented Thinking Throughout the Development Process

Whereas we considered the idea of a backlog as an evolution to a project plan, we may further consider a strategic roadmap as an evolution to a backlog. We discussed that a backlog does not usually scale to products or services of higher complexity since they would become cognitively unmanageable. In our current world, however, products and services steadily grow in their complexity as, for example, compared to the very first smartphone apps. This, in turn, demands new tools for practitioners.

However, a roadmap is not necessarily to be seen as a complete replacement for a backlog. It is also reasonable and pragmatic to use the two in conjunction with each other, especially on different levels of granularity and for different purposes. The roadmap then dedicatedly serves the purpose of strategic prioritization while the backlog is used for steering work packages on the operational level (see Fig. 2).

Our notion of a well-designed strategic roadmap comes along with distinctive features, which are described in more detail below.

Communication and Alignment Between Stakeholders

The roadmap is meant as a communication artifact to inform and align stakeholders on slicing the individual delivery increments of the product or service. A delivery increment is a set of features that will be released together on a defined milestone (see Fig. 3). This set of features ideally delivers recognizable value to its customers

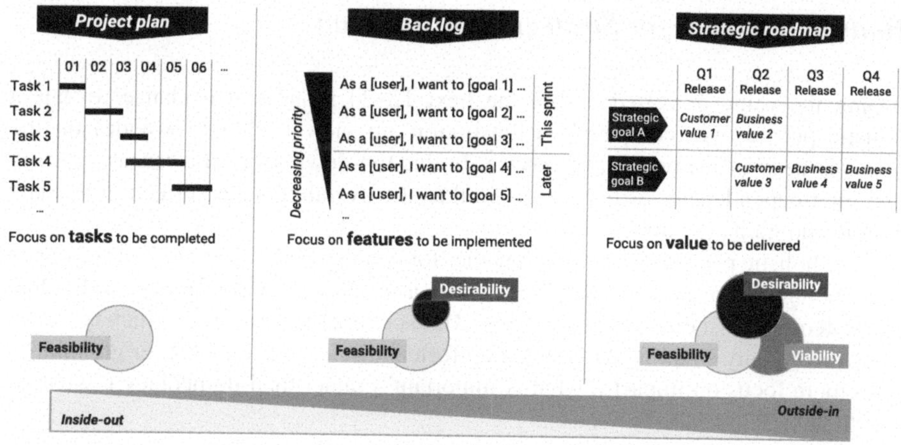

Fig. 2 Comparison between project plan, backlog, and strategic roadmap

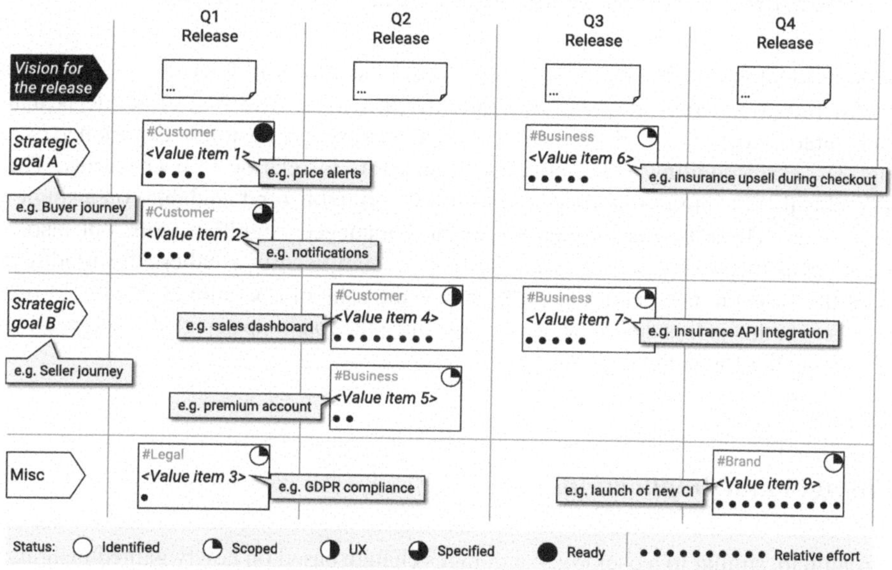

Fig. 3 Structure of a strategic roadmap

(w.r.t. desirability) or the company itself (w.r.t. viability) to avoid the watering can effect described above. Therefore, the underlying structure of the roadmap is a time grid, which corresponds to the release cycle of the product or service. This may, for example, be a quarterly, monthly, or bi-weekly grid used to define what scope will be delivered in which release. Whereas backlogs usually plan ahead for just the next 2–5 sprints (2–3 weeks each, that is, no longer than 15 weeks), roadmaps typically cover the next 12 to 24 months.

Holistic Overview for Strategic Prioritization

Seeing the value delivered within the next 12–24 months at a glance reveals a holistic perspective (compared to the bottom-up view backlogs provide) on the value strategy of the product or service. In this top-down view, the causal structure and inter-dependencies of features as well as the overall strategic direction become visible and guide the discussions.

This holistic perspective is a key enabler for strategic prioritization. Not only does it encourage value-oriented thinking, but it also allows for prioritizing individual value items against one another (see Fig. 3). In contrast to a backlog, which typically consists of many small items, a roadmap fosters discussions on a higher granularity with more focus on impact ("what is important") rather than immediacy ("what do we do next").

Prioritization Through Meaningful Levels of Granularity

In order to hit the "right" granularity (i.e., level of abstraction) for value items on the roadmap, we generally propose two guiding principles. The first principle is that each value item on the roadmap is framed as a macroscopic customer experience (see Fig. 3)—corresponding to the concept of an EPIC in relation to user stories. The second is that the entire roadmap should still be printable (and readable) on a single-page sheet. These two principles allow for a managed cognitive load, but also a meaningful discussion. If the granularity gets significantly smaller, the roadmap loses the strategic perspective it provides for the sake of operational details. In our experience, this is less desirable, especially since the roadmap serves the perspective of seeing "the forest through the trees."

Changes and Implications

A roadmap, similar to a backlog, encourages change based on newly gained insights. In contrast to a project plan, a roadmap is not a fixed plan, but rather a snapshot of the current discussions around the product or service. Since value items are ideally framed on a similar granularity, they become interchangeable in their priority, respectively implementation order (at least unless causal dependencies prevent this). Discussing a re-prioritization along the roadmap allows understanding its implications in the context of: "If we shift feature A to the next release for the sake of feature B, we will delay our expected revenue growth by 3 months. Can we afford to do this?" If a valuable item is prioritized higher, another value item needs to be prioritized lower to compensate for the effect, assuming that the implementation capacity remains the same. This way, different stakeholder interests can be made explicit and negotiated on a tangible discussion ground while the overall integrity of the product or service remains in sight.

Based on our industry observations, we do not conclude that better tools necessarily produce better results. Framing roadmap items that actually deliver value and hitting the right granularity are tasks that cannot be taken over by tools. These still heavily depend on the practice itself. We ourselves often use a custom-designed sheet and bring a printed copy for every stakeholder at the meeting. That way, everyone involved can pick up a pencil and suggest alterations to it, rather than getting lost in an expert tool that only a few people involved can practically use.

Conclusion

The notion of value-oriented thinking is already deeply rooted within many startups since the need to deliver value is tightly bound to the survival of their existence as a company. A startup simply cannot afford to not deliver value to their customers. Hence, keeping a close eye on the strategic prioritization of value to be delivered is vital. It explains, why startups use strategic roadmaps for the purpose of facilitating discussions both internally and with external stakeholders (e.g., investors). In conclusion, strategic roadmaps serve as an essential means to minimize existential risks for startups.

In the corporate world, however, we have encountered far fewer strategic roadmaps in practice so far. This may be caused by (1) the fact that project plans and backlogs are already manifested as common practice there, (2) this practice is not actively reflected upon or challenged, and (3) that a strict value-oriented thinking might not be as existential for the entire corporation as it is for a startup. Corporations may potentially be more likely to compromise on the value perspective in order to better manage feasibility risks. Nevertheless, we still consider it beneficial to utilize strategic roadmaps in the context of a product or service innovation. The value discussion then becomes explicit, as in Design Thinking, and an outside-in mindset can be maintained more easily throughout the requirements engineering and subsequent stages in the development phases as well.

While we have primarily discussed the limits of project plans and backlogs and how strategic roadmaps can improve strategic prioritization, we do not wish to downplay the benefits of project plans or backlogs in general. On the contrary, these two artifacts also have their fitting place, and each serves a meaningful purpose. At the same time, we do not consider them a sufficient means to facilitate an ongoing value-oriented strategic prioritization in the context of developing product or service innovations.

Feel free to contact the authors for a template on strategic roadmaps.

References

Brown T (2008) Design thinking. Harv Bus Rev 86:84–92

Green C (2014) The silent killer of new products: lazy pricing. Harv Bus Rev. https://hbr.org/2014/09/the-silent-killer-of-new-products-lazy-pricing. Accessed 19 July 2021

Kelley T, Kelley D (2013) Creative confidence. Unleashing the creative potential within us. Crown Business, New York

Moore GA (2014) Crossing the chasm: marketing and selling technology products to mainstream customers. Harper Business, New York

Schmiedgen J, Rhinow H, Köppen E, Meinel C (2015) Parts without a whole? The current state of design thinking practice in organizations. Universitätsverlag, Potsdam

Sutherland J (2014) Scrum: the art of doing twice the work in half the time. Penguin LCC US

Managing Tensions in Research Consortia with Design Thinking Artifacts

Dario Staehelin, Mateusz Dolata, and Gerhard Schwabe

Introduction

Innovation projects between industry and universities are popular. The industry partners receive access to the most recent scholarship, while the researchers benefit from easier access to the field, supporting the external validity and the relevance of their research. Additionally, the resulting third-party funds allow more researchers to pursue their careers in academia. The Information Systems (IS) discipline values collaborations oriented at the development and evaluation of new systems or technologies to support the practitioners. However, innovation projects like these are inevitably related to tensions resulting from differences in culture and incentive systems between academia and industry, as well as among multiple industrial or research partners.

Whereas some level of tension was shown to spark creativity (Badke-Schaub et al. 2010), too much tension might be detrimental (Dolata and Schwabe 2014). It might generate interpersonal and interorganizational conflicts and turn a visionary project into a search for the least common denominator. Researchers claim that Design Thinking maintains a healthy level of tension in creative teams by offering adequate processes, techniques, as well as a general user-centered mindset (Leifer and Steinert 2014). The core of Design Thinking is the iterative development and testing of prototypes with relevant stakeholders. Apart from prototypes, Design Thinking projects frequently create further artifacts that are not necessarily developed for evaluation (e.g., meeting reports). Previous research on Design Thinking in engineering education confirms the role of Design Thinking artifacts for the maintenance of shared models among engineering students (Edelman et al. 2009). There is evidence confirming the role of Design Thinking artifacts in identifying and

D. Staehelin (✉) · M. Dolata · G. Schwabe
Department of Informatics, University of Zurich, Zürich, Switzerland
e-mail: staehelin@ifi.uzh.ch; dolata@ifi.uzh.ch; schwabe@ifi.uzh.ch

J. Hehn et al. (eds.), *Design Thinking for Software Engineering*, Progress in IS,
https://doi.org/10.1007/978-3-030-90594-1_9

maintaining the compatibility of software requirements in Design Thinking-based consulting (Hehn et al. 2020). However, little attention was paid to innovation projects between academia and industry (Dolata and Schwabe 2016) and, specifically, to the role of artifacts in such innovation projects. Different than in student or consulting projects, various visions, objectives, and work cultures in interorganizational and cross-disciplinary projects might generate more intense tensions. It remains unclear if, to what extent, and how Design Thinking artifacts can contribute towards tension management and effective collaboration in those projects as well. Thus, in this contribution, we explore *how Design Thinking artifacts can be used to manage tensions in research consortia.*

This contribution is a single-case study zooming in on an ongoing project, "SpeechAdvice." Four industry and two academic partners formed a research consortium. The project was established to pursue a vision of a conversational agent for face-to-face financial counseling. The contribution engages in an analysis of the six-month exploration part of the project in which the abstract vision should have been transferred in a set of functional and non-functional requirements to be used in the subsequent development-evaluation iterations. The reflection shows how various artifacts developed by the project participants transfer meanings across boundaries, embody knowledge, and facilitate interaction and direct manipulation. The artifacts and interaction with the artifacts were shown to support the identification, explication, and maintenance of the tensions between the project participants. While our case study focused on a research consortium, we assume the findings to be transferable to interdisciplinary and interorganizational projects without participation from research institutions.

Related Work

The term *Design Thinking* describes a family of approaches that propose the use of product designers' practices to innovation. Design Thinking originates from academia, where it was used for teaching mechanical engineering (Carleton and Leifer 2009; Johansson-Sköldberg et al. 2013). Approaches inspired by this paradigm found their way to industry and were applied to solve practical problems (Johansson-Sköldberg et al. 2013; Brown 2008). Successes of consulting firms like IDEO popularized this methodology. Design Thinking was applied in software development (Lindberg et al. 2011, 2012), service innovation (Ojasalo et al. 2015), and business process innovation (Luebbe and Weske 2012). Despite the academic origin of Design Thinking, reports on the application of Design Thinking in collaboration between university researchers and industrial stakeholders remain very rare (Dolata and Schwabe 2016) and mostly attend to situations where Design Thinking researchers help establish a Design Thinking-oriented innovation culture in a company (Vetterli et al. 2012). The rare reports on the use of Design Thinking in design-based research might be related to Design Thinking's focus on innovation rather than scientific rigor, unlike other acknowledged paradigms, e.g., design science research

(Hevner et al. 2004). However, some accounts suggesting that Design Thinking is compatible with design-oriented research and, thus, suits the collaboration between academia and industry who join forces in an innovation project (Dolata and Schwabe 2016). We speculate that Design Thinking can be beneficial for such collaborations because of its explicit focus on maintaining contradictions between various ideas or opinions and its orientation towards artifacts as means of collaboration.

Research and innovation processes both involve certain levels of contradiction. Both need to handle contradictory evidence, differing preferences, and conflicting lines of reasoning. Of course, they differ in how they value rigor, relevance, and usefulness of insights and developed diverse ways to maintain tensions that emerge along the process. In a collaboration between research-oriented institutions and innovation-oriented industries, the emergence of tensions is natural (Nunamaker et al. 2015). Typically, tensions are conflicts of opposing mostly independent interests. Organizational literature regards tensions as conflicting demands that can only be eliminated through tradeoffs (Smith et al. 2017). With this increased complexity, organizational research began viewing tensions as interdependent demands enabling instead of hindering change (Poole and Van de Ven 1989). This emerging view is that tensions do not necessarily need to be remediated or resolved but rather that they need to be managed appropriately. Design Thinking provides a variety of methods to use tensions to spark creativity (Dolata and Schwabe 2016). However, it remains underexplored how and to what extent those methods are sufficient to manage tensions in projects between researchers and industry stakeholders.

Among others, Design Thinking suggests the use of prototypes and other tangible artifacts as the primary objects of all activities. This aligns with materialistic, socio-material, and ecological perspectives on work and human activity (Gibson 1979a, b; Orlikowski 2010). Accordingly, objects co-define the activity at hand by providing its direction, motivation, and meaning (Leontyev 2009; Karanasios and Allen 2018). The role of artifacts for innovation and inventions has been widely discussed concerning capturing, transfer, generation, and broadcasting of knowledge in intraorganizational innovation (Ciriello et al. 2014) and cross-disciplinary collaboration settings (Nicolini et al. 2012). The research has identified four categories of artifacts: activity objects, boundary objects, epistemic objects, and infrastructure objects. Activity objects are artifacts that motivate collaboration and manipulation among involved individuals (Nicolini et al. 2012). Boundary objects facilitate translation and transformation across disciplinary or organizational contexts, thus acting as enablers of collaboration across boundaries (Ciriello et al. 2014; Nicolini et al. 2012). Epistemic objects raise curiosity about the known and unknown, thus acquiring emotional, social, or intellectual engagement from its creators (Ciriello et al. 2014; Nicolini et al. 2012). Finally, infrastructure objects are artifacts that enable and support interaction between individuals (Nicolini et al. 2012). Studies using this classification could show that, based on their affordances, artifacts in innovation projects belong to a specific category.

However, this classification emerged as an outcome of reflection upon research and engineering projects and not in relation to Design Thinking. Whereas many innovation projects generate multiple artifacts in the process of creating and managing knowledge and insight, Design Thinking demands the production of artifacts as a way to create something new in the form of prototypes (Uebernickel and Brenner 2016). To make it more explicit: whereas in most innovation projects, artifacts emerge along the process, Design Thinking defines its milestones through artifacts to be produced. Those artifacts should be subsequently used for the collection of feedback and manipulation. This suggests that the Design Thinking artifacts would primarily act as activity objects. But is it that simple? Previous research shows that engineering student teams use prototypes and other artifacts to align their shared mental models (Leifer and Steinert 2011). By-products in student teams can even deal as measurement instruments for progress and performance in Design Thinking (Dolata et al. 2017). Similarly, Design Thinking artifacts support the translation of field insights obtained from practitioners into requirements useful for developers (Hehn et al. 2020). This suggests that Design Thinking artifacts can take on various roles; however a systematic observation of how Design Thinking artifacts fulfill those roles is missing.

Case Description

Financial advice is becoming increasingly complex. Client advisors are required to provide exceptional customer services, recommend the best products, consider potential risks, identify cross-selling opportunities, and keep regulatory-relevant protocols, all while developing and maintaining a sustainable relationship with the client (Dolata and Schwabe 2017a, b). All these tasks increase the cognitive load on the advisor, which may decrease the quality of the collaboration and communication between the client and advisor (Schwabe and Nussbaumer 2009; Kilic et al. 2017).

Artificial intelligence (AI) bears the potential to support advisory sessions by performing tasks for both the advisor and the client. The client advisor can focus on his core competency—building relationships with her customers—by seamlessly integrating a digital assistant that carries out time-consuming tasks in the background (e.g., looking up a stock chart or filling in a legal document) (Dolata et al. 2019). Optimally, such a digital assistant would not only be operated by command and control and provide information on simple queries—like existing digital assistants as Siri—but also be able to interact autonomously with the advisor and the client and handle more complex tasks like risk evaluations of investment decisions.

Funded by InnoSuisse—Swiss Innovation Agency, a project consortium set out to explore this vision of AI-enhanced bank advice in the project "SpeechAdvice." The following six partners belong to this consortium:

- The Swiss cantonal bank "SwissCantonal" and the Swiss regional bank "SwissRegional" are interested in supporting advisory sessions with the latest technology to create an excellent experience for the client and reduce the cognitive load on their client advisors. Within the project, both banks are responsible to provide the project team with access to the field (i.e., client advisors and clients) and guaranteeing that the developed solution aligns with the long-term strategies of the banks. They hope that the integration of AI in financial advice increases efficiency and attracts new clients.

- The small IT company "SpeechExcellence" focuses on signal processing and speech recognition. Its technology enables the development of automatic speech recognition systems, which is a crucial part of the technology underlying the vision of the digital assistant. Within the project, SpeechExcellence is responsible for speech recognition in Swiss German. They are interested in developing and marketing a showcase for their technology.
- The larger IT company "EngineeringExcellence" has substantial experience in the implementation of digital trends as software solutions. Within the project, EngineeringExcellence is responsible for the main development of the digital assistant. They intend to acquire new capabilities concerning the development of AI products during the project and see it as a chance to develop a product to be marketed in the service sector.
- The Institute of Interactive Technologies (IIT) of a university of applied sciences has substantial experience with interactive technologies, including natural user interfaces and extended reality. Within the project, IIT is responsible for information visualization and augmented reality. Their interest lays in advancing multi-modal interaction concepts between humans and technology.
- The Information Management Research Group (IMRG) of the Department of Informatics at the University of Zurich has extensively researched the potential of collaborative IT systems to enhance advisory quality. Within the project, IMRG is responsible for the exploration and definition of a use case for the conversational agent (CA) by applying design science research. They are interested in generating knowledge in the field of machines as teammates and seek to generalize their findings for the scientific community.

Defining a Use Case with the Double Diamond Process

SpeechAdvice is a technology-driven project: All partners are interested in how advances in speech recognition, language processing, and machine learning may change financial advice-giving. Note that such a process is quite typical for companies "exploring on the edge" (Sambamurthy and Zmud 2012) but at odds with typical problem-driven design research (Peffers et al. 2007). Design thinking also typically prioritizes customer needs to technology potential (Brown 2008). So, the project's exploration part aims to transform SpeechAdvice into a problem-driven project by identifying and defining a use case for the conversational agent in financial advice. To achieve this, the project team adopted the double diamond process (DDP, see Fig. 1). Unlike typical Design Thinking projects, the consortium already identified a specific technology as the center of the solution.

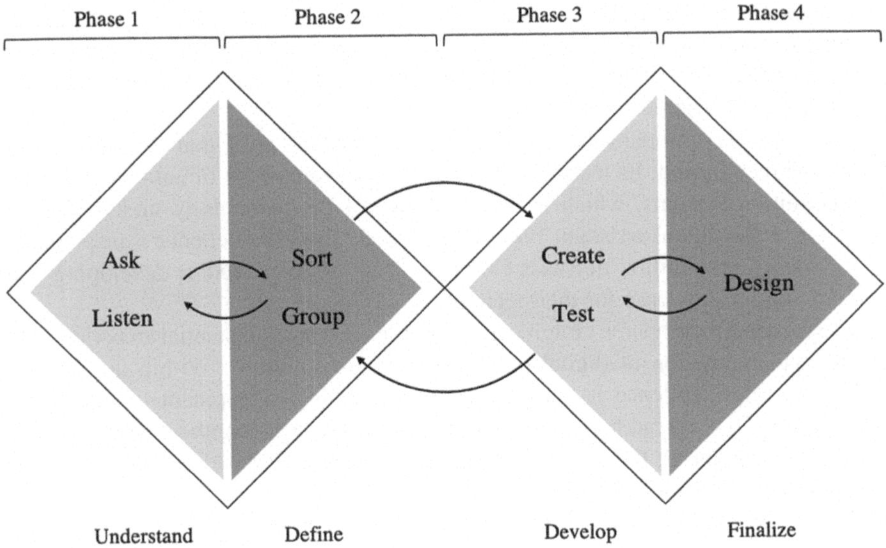

Fig. 1 Double Diamond process

Understand In the first phase, the research consortium aimed at accurately understanding the problem of financial advice. For this purpose, the project team acquired an extensive list of current problems by *asking* client advisors about their daily troubles and *listening* to stakeholders included in the project.

Define The research consortium *sorted* the problems along the customer journey. They *grouped* similar problems reducing the list to meta problems which were then rated according to their relevance and suitability for the SpeechAdvice project.

Develop Entering the development phase, the research consortium explored a variety of solutions to the most promising problems. Iteratively, they *created* and *tested* prototypes in close collaboration with the end-user.

Finalize Converging a second time, the research consortium selected the use case and *designed* a high-fidelity prototype. A final user evaluation concluded the Double Diamond Process.

Model Development

During the exploration part of the "SpeechAdvice" project, many conflicts arose as symptoms of underlying tensions. The goal of the exploration part was to establish a use case to be realized in the subsequent implementation part (while leaving a margin of flexibility). While the tensions could be used to identify various problems and generate diverse ideas, they needed to be addressed to successfully define a use

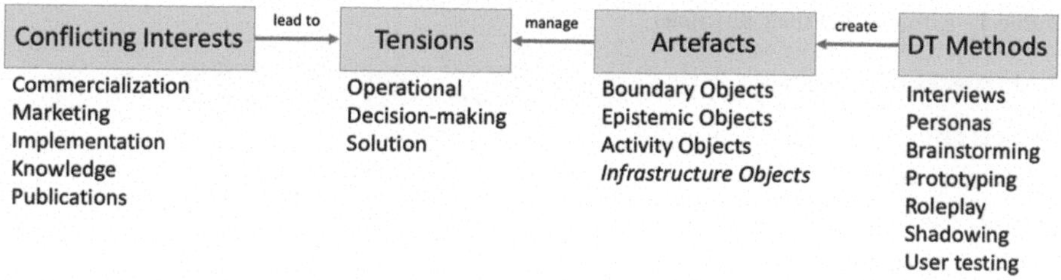

Fig. 2 Tension model

case while considering the diverging interests of the consortium members. Our tension model (Fig. 2) gives an over the relationship between conflicting interests, tensions, artifacts, and design thinking methods.

Conflicting Interests The interest in advancing technology-supported advisory services formed the common ground for the project consortium. Especially, the idea to develop a CA to reduce the complexity of a service encounter was of specific interest to all parties. However, the driving forces within the organizations are diverging fundamentally. On the one hand, the banks and both IT companies have mostly commercial interests in the project. The IT companies intend to develop a solution that creates value for their clients. Both banks expect higher efficiency (i.e., reduced workload for the advisory) and marketing potential to attract new or retain existing customers. In both cases, the primary goal is to gain a competitive advantage over their competitors as fast as possible while at the same time minimizing their financial risks. On the other hand, IIT and IMRG are especially interested in advancing knowledge in their respective fields and making their findings accessible to the public through publications and presentations. This requires a systematic approach to understand problems, their root causes, and possible solutions. Applying methodologies, collecting data, and relating insights to the state of the Art of academic literature requires time and financial resources. One can boil down this conflict in two statements: EngineeringExcellence wanted to "Fail often, fail fast" while IMRG wanted to "Fail - based on deep insights." These conflicting interests became the main driver for the tensions between the consortium members.

Tensions Tensions are the results of conflicting interests. They manifest in the behavior of the individual actors. In the exploratory phase, we encountered three major tensions:

- *Operational tensions* manifest themselves in different approaches on the operationalization of the project plan and understandings of the applied methods.
- *Decision-making tensions* become apparent in contrasting approaches to reach decisions.
- *Solution tensions* surface in opposing ideas about the novelty of the intended solution.

Table 1 Artifacts and their functions

Artifact	Boundary Object	Epistemic Object	Activity Object
Project plan	X		
Problem scenario	X		(X)
Activity scenario		X	(X)
Information scenario		X	(X)
Job description		X	(X)
Prototypes	(X)	(X)	X
Functionalities map	X		

X = primary function, (X) = secondary function

Design Thinking Methods Design Thinking methods were applied as a methodology to manage tensions as part of user-centric IT development. The consortium agreed that a strong involvement of end-users (i.e., client advisors) would be necessary to understand their needs and problems better. Design Thinking should guide the project team towards a relevant use case. Furthermore, the project team applied Design Thinking as a toolset: Interviews, prototyping, role play, shadowing, and user testing helped bridge industry and research objectives. The scenario-based design framework (Rosson and Carroll 2009) provided additional guidance. Design Thinking, in combination with scenario-based design produced many artifacts facilitating the management of tensions.

Artifact Artifacts are the outcome of the application of Design Thinking methods. They helped to resolve tensions by serving as a boundary object, epistemic objects, or activity objects. Material infrastructures are not highlighted in this paper as they did not affect the tensions in this project. Table 1 lists the relevant artifacts of the project and their functions:

Analysis

For the analysis, the author group reviewed the data described above (i.e., interviews transcriptions, workshop materials) complemented with meeting minutes. In several workshops, the authors identified sequences of conflicts and their origins. In several iterations, the authors condensed the sequences into tensions. In the following, we illustrate the three tensions with episodes from the project. Then we analyze the role of artifacts in managing these tensions. Table 2 summarises the key drivers of the stakeholder in each tension.

Operational Tensions The following episode took place during the workshop "Creating a customer journey." The goal of this workshop was to understand the

Table 2 Drivers of the stakeholders for each tension

Tension/ Stakeholders	Operational Tension	Decision-making Tension	Solution Tension
SwissCantonal/ SwissRegional	User-oriented	Experience-driven	Commercial/ Marketing interests
SpeechExcellence	Solution-oriented	n/a	Commercial interests
EngineeringExcellence	Solution-oriented	Experience-driven	Commercial interests
IIT	Exploration-oriented	Facts-driven	Interest in innovation
IMRG	Exploration-oriented	Facts-driven	Interest in innovation

current problems of bank advisory services from the standpoint of the clients and advisors. The participants were asked to brainstorm problems in interdisciplinary groups. The identified problems should then be sorted along the customer journey on a large printout and discussed. Afterward, the groups joined the plenum and presented and discussed their outcomes.

Sarah, Christine, Bill, and Tom participated in the workshop lead by George and formed a breakout group. At this point, Tom and Christine (working for IMRG and IIT) have conducted interviews with client advisors and Tom already observed two advisory sessions. Sarah represented the banks as she is a former advisor now working as an instructor for junior advisors. Bill is a software engineer working for EngineeringExcellence.

In the beginning, George, the workshop leader, highlighted the focus of the workshop. He told the participants about the interviews and observations and their preliminary results. After this introduction, the breakout group started brainstorming and writing problems on Post-it notes. They placed the notes on a printed customer journey after shortly explaining the problems. After a while, Christine observed that she was the only one brainstorming problems of the clients. She told Tom her observation and he agreed. He also observed that most problems were focused on technical aspects of the advisory session disregarding the client's initial contact and the advisor's preparation and postprocessing. Bill and Sarah argued that the technical solution will probably support the actual advisory session. Therefore, they should focus on problems during the service encounter due to the restricted time and resources of the project. Christine and Tom disagreed and referred to the interviews where the advisors mentioned a lot of potential before and after the advisory session. Additionally, they emphasized the need to first understand problems in a broader sense. It became clear, that the group had a different understanding of the problem and solution scope. They agreed to discuss this matter in the plenum after the breakout session as they could not reach a consensus.

In the plenum, the disagreement proved to affect all other groups as well leading to discussions about different interpretations of this workshop's goal. While the representatives from the bank focused on the advisors' overall contact with the clients, the IT companies emphasized the technical limitations of the technology used during the advisory session. IIT and IMRG applied a broader view to understand the problems of both parties across the whole journey. George used the project plan to explain the four phases and their purpose to the participants. The technology partners and bank representatives realized that a broader mindset would help them understand the roles of clients and advisors better. Hence, they will be able to design a better solution supporting both parties. Additionally, George illustrated how the results of the workshop form the basis of the problem scenarios. In the following, the teams collaboratively created an abundance of problems for both, the client and advisor, considering all phases of the journey.

After the workshop, the conflicting views were rejoined by the problem scenarios. The project team discussed their interests and the current shortcomings. Bill highlighted several sections where he was missing banking knowledge. Generally, EngineeringExcellence experienced great difficulty with working on and elaborate problem definition as the solution—the CA—appeared to be clear from the start. As a result, they urged to identify problems where a CA would be an obvious solution. On the other hand, IIT emphasized the need to further understand the clients' problems as no interviews with clients were conducted so far. The problem scenarios enabled the team to incorporate their different interests and align their understanding.

In this episode, three tendencies could be observed: Firstly, the banks claimed that the focus should lie on the bank advisor. Secondly, the technology partner focused on problems where a CA was a feasible solution. Lastly, the research partners tried to balance the views and maintain an open mind towards less obvious problems. The project plan and problem scenarios functioned as *boundary objects* to align the conflicting views by establishing a common language. They facilitated clear communication across disciplines. Even though the project partners agreed on the overall project approach, the first workshop revealed conflicts when translating into practice. In the scenario above, George used the project plan as a *boundary object* to address the conflict and enable collaboration between the participants. In this situation, the project plan worked as a translator between the disciplines. Regardless of the participants' agreement with the choice of the methods, they recognized the different interpretations of the goal and aligned their efforts. The problem scenarios played a very similar role. While the project plan aligned the understanding of the applied methods, the problem scenario had the same effect on the problems identified as relevant. The project team explicated their interpretation of the problem by writing it down. Misunderstandings and diverging interests explicated through discussions and formed the problem scenarios until a mutual agreement was achieved.

Decision-Making Tensions After understanding and defining relevant problems, the project team moved on to the development phase. In a workshop, participants were asked to create solutions to the selected problems by "Enacting the Future." They were asked to first rapid prototype material their CA would need. Afterward, the groups should enact a future advisory session using their prototypes. After presenting their advisory session in the plenum, the participants would discuss the different solutions.

Eva, Barbara, Peter, and Joe formed a breakout group and started brainstorming on functional aspects of the CA. *Peter, experienced in human-computer interaction, drew screens on paper according to the input of the others. After the group created a set of screens consisting of different charts and portfolio designs, they moved on to the physical representation of the* CA. *Barbara, Peter, and Joe actively described their vision of the CA and its capabilities to interact with the client and advisor through gestures and facial expressions. They imagined some kind of anthropomorphic character, maybe the banks' mascot if they have one. However, Eva, working in the IT department of one of the banks, said that CA should not have a physical representation. In her opinion, the advisors would not welcome a CA with a physical representation other than something similar to an Amazon Echo. Barbara insisted that research has shown that an anthropomorphic representation creates a social attachment between the user and the system. Peter supported Barbara's argument while Joe, on the other side, agreed with Eva. Joe's prior experience with CA was rather negative and he believes that a physical representation could raise concerns about the CA with the clients. The two sides argued intensely whether or not the CA should have a physical representation. Facts, presented by Barbara and Peter could not convince Eva and Joe. Eva strongly believed that research in other fields could not be transferred to the financial industry while Joe questioned the transfer into a real-life setting. With time running short, they settled for a chess piece as a physical representation of the CA for the time being. However, the dispute would arise again after the workshop.*

The data screens, physical representations, and enacted advisory sessions were transferred into activity scenarios. The project team discussed the results in one of their regular meetings. Again, the conflict on the physical representation arose. Still, Eva argued against an anthropomorphic representation based on her opinion. However, Joe's concerns were partially resolved after he looked at the other group's creations. He started to engage with the prototypes and envision how the CA could use even more channels to interact with the users. Nevertheless, the unknown opinions of the advisors delayed a decision and the argument could not be settled. Finally, the project team decided after the research partners interviewed advisors.

The conflict described in the episode above arose from different approaches in the decision-making process. Firstly, the banks based their arguments upon prior experience with advisors when introducing changes in their work process. Independent from organizations, others were led by their own experience with CA (e.g., chatbots in customer services). In contrast, the research partners asked for facts as a basis for decisions. Consequently, this conflict hindered collaboration as decision-making was slow.

Activity scenarios and prototypes facilitated managing the *Decision-making tensions*. Especially considered as *epistemic objects*, these artifacts enabled the project team to differentiate between opinions and facts. Enhancing the written scenarios with illustrations allowed the team to question decisions and identify potential knowledge gaps. As a result, both opinions and facts were included in the activity scenario. Validating assumptions advanced the knowledge of potential applications of a CA in bank advisory. Furthermore, prototypes also functioned as an *epistemic object* in this context. Brainstorming different types of physical representation affected people's beliefs. As seen above, Joe realized the potential of an anthropomorphic CA only after being confronted with the prototypes. In the following, the team members referred to specific prototypes to express their ideas. The artifacts seemed to stimulate the team and triggered a growing attachment not only to the prototype itself but to the emerging solutions and the overall project. As a secondary effect, the activity scenarios and prototypes guided the activities of the project. The conflicts inherent to the artifacts triggered creative refinement of the solution as the contradictions were addressed collaboratively. The information scenario revealed itself as a powerful *activity object*. As a final form of the use case description, the information scenario contained some conflicts that triggered additional artifacts like the job description guiding group work and generating knowledge by resolving the conflicts.

Solution Tensions Before finalizing the use case for the CA, the project team met for a final workshop. The vision has already been reviewed by an expert panel and six bank advisors. Therefore, the goal of this final workshop was not to challenge the general idea but to elaborate on the requirements of the solution. The workshop was again led by George and the eight members mentioned above participated. The information scenario, prototypes, and job description built the foundation for the workshop.

After George introduced the participants to the workshop by summarizing the use case, the participants were asked to write down short stories. The goal of this exercise was to imagine how they, as a CA, would perform a certain task. For this, the participants should consider both technical and non-technical aspects. For each story, the group then discussed aspects of the CA that they would like to keep, omit, or add. In one group, Bill's concerns about the technical feasibility surfaced again. He was anxious about the high expectations the others had towards the CA. *In contrast to the stories of his group members, his story was focused on enhancing the current practice rather than trying to advance advice-giving. In contrast, Barbara came up with novel ideas about how the CA influence the course of the session with a certain autonomy. Sarah's idea went in the same direction. In the following, they discussed whether or not the functionalities should be restricted by the technical level of difficulty. Barbara insisted that novelty is a major aspect as they would participate in a research project. Representing the technology partners, Bill and Joe argued that too high expectations could potentially doom the project. As a result, the group presented a conflicting list of feedback to functionalities to the plenum. As it turned out, the other groups experienced similar conflicts. One part of the participants favored evolving existing concepts in bank advisory, while the other part rethought these patterns by inventing new patterns. For example, Tom envisioned the CA to take an independent role in the advisory session instead of simply assisting the client advisor. Such a non-functional requirement was immediately challenged by Bill and Joe as this would be hardly feasible with available frameworks.*

Due to the limited time, the conflict had to be addressed after the workshop. The technology partners were worried about the technical feasibility. They strongly argued against an overly ambitious vision. Whereas the research partners were not satisfied with an out-of-the-box solution that may increases efficiency but would not question existing practices. The banks, in between the two fronts, weighed commercial against marketing objectives. While an innovative solution might increase their revenues, a true invention could boost their market share and attract new clients. The project members agreed on a staggered approach prioritizing the functionalities by the level of technical difficulty. For this, a high-fidelity prototype was developed to illustrate the vision. Simultaneously, the project team created a functionalities map including dependencies between functionalities. As a result, both, the concerns of the technology partners and the wish for an inventive solution were considered.

This episode illustrates the origin of the *Solution tensions*.

The consortium members had a different understanding of the target solution and its novelty. Both technology partners were driven by commercial interests: SpeechExcellence intended to demonstrate the capabilities of their speech recognition in Swiss German in a working product; EngineeringExcellence expects to foster their connection to the financial industry with a new product offering. On the contrary, both research partners strive to generate knowledge in their fields. IIT was interested in expanding interactions between humans and technology. IMRG researches the implication of a CA on the dynamic of advisory sessions.

The functionalities map and high-fidelity prototype were mainly used by the project team to address the *Solution tensions*. Initially, the functionalities map and prototype explicated the different perceptions of novelty and formed the ground to overcome the conflicting interests. The project team could agree on a common understanding of the scope of the prototype as they listed the functionalities required for the use case and their dependencies. Consequently, both artifacts as *boundary objects* made the continuance of the collaboration possible. Once the understandings were aligned, the Prototype (taking the role of an *activity object*) directed the activities necessary to realize a first idea of the final solution. The project team identified conflicts in functional and non-functional requirements they failed to notice in the information scenario. Finally, they reached a solution the whole consortium desired to develop.

Implications and Conclusion

This contribution discusses the effect of artifacts in managing tensions in interdisciplinary Design Thinking projects. It contributes to Design Thinking literature by explicating the importance of artifacts for collaboration between research and industry. Additionally, we shed light on the mechanics of managing tensions with artifacts by applying the framework proposed by Nicolini et al. (2012).

Our study amends existing literature on Design Thinking by utilizing it in innovation-oriented collaboration between academia and industry, thus going beyond consulting and purely academic applications (Hehn et al. 2020; Brown 2008; Vetterli et al. 2012). Additionally, our study confirms the suitability of Design Thinking in technology-driven research. Our results show that techniques proposed by Design Thinking needed adaptation to the particular context. For instance, the rigor aspects relevant for academic institutions required considerations along the process. Academic rigor requires the understanding of theoretical and empirical underpinnings for major design decisions and capturing their rationale rather than relying purely on intuition and experience. Adapting Design Thinking methods has externalized existing or created new tensions because they required more explicit communication between the stakeholders. Overall, Design Thinking turned out to be effective in managing tensions by relocating them from interpersonal conversation to external artifacts. As a result, conflicts were linked less to a particular person or organization but to an artifact that simplified criticism and let individuals distance themselves from their idea. Furthermore, artifacts played a crucial role in the identification of diverging interests and conflicts and the appropriate handling of these. The material representation could capture the contradictory opinions and motivate further engagement with the project.

Additionally, our study contributes to the socio-material and materialistic approaches towards innovation. It applies the categorization proposed earlier (Ciriello et al. 2014; Nicolini et al. 2012) to new categories of projects, namely, Design Thinking-driven projects. Whereas previous artifact-oriented analyses of innovation processes tended to classify artifacts into single categories based on their affordances, this study suggests an adapting role of the same artifact depending on the specific context. This is, in essence, not contradictory with the affordance and ecological theories stating that affordances of an object emerge not only based on individual features of the object but also based on the context defined by the surrounding objects and their characteristics (Nicolini et al. 2012). Specifically, in a conflict situation, a prototype, which previously acted as an activity object, changes its character to a boundary object (to translate the meaning across disciplinary boundaries). This extends the discourse on the roles of objects in innovation (Ciriello et al. 2014; Nicolini et al. 2012).

Those insights have practical relevance for Design Thinking facilitators as well as collaboration between academia and industry. Design Thinking prescribes the use of various artifacts but frequently limits itself to explaining their role in relation to the content of a project, i.e., an idea, a vision, or an ultimate product. This is reflected by the naming of these artifacts—they are called prototypes. Our reflection makes clear that the artifacts simultaneously take on further roles, in particular, they help managing tensions within the team. Design Thinking facilitators might use this insight and suggest specific forms of prototypes to bring some tensions to the surface or to help the team resolve conflicts. Given the variety of methods and prototypes counted towards Design Thinking, facilitators often face the problem of whether one or the other is more appropriate and what factors should be considered to assess the appropriateness. We claim that considering the tensions in the team and the role which the artifact should play in the subsequent collaboration (boundary, activity, or epistemic object) might support the decision in this regard. This also suggests that there is no "one fits all" sequence of Design Thinking methods or prototypes: the individual situation of the team needs to be considered when deciding about the subsequent steps. This strengthens the position of Design Thinking facilitators.

Collaboration between academia and industry is essential for the effective application of scientific progress in practice. However, there is little guidance on how to set up and conduct such projects, given the inherent tensions and conflicts. This contribution makes clear that the material aspect of the collaboration is key to specifying the differences and resolving them. At the same time, the developed artifacts form a materialized memory of the project allowing for reconstruction of the decisions taken on the go. This allows for the rigorous presentation of the project and its results, which is important for academia.

In conclusion, our study broadens the view on artifacts created by applying Design Thinking and highlights their capabilities in managing tensions in cross-disciplinary collaboration. We emphasize the importance of consciously choosing Design Thinking methods to create artifacts that direct activities. Furthermore, we introduce the multidimensional framework by Nicolini et al. (2012) to Design Thinking, exploring the potential of artifacts to bridge boundaries and trigger cross-disciplinary collaboration.

For practitioners, this study offers several implications. Firstly, explicating organizational and personal interests at an early stage discloses potential tensions. This enables the project leader to address these tensions by selecting appropriate Design Thinking methods proactively. Secondly, combining typical Design Thinking methods with other user-centric design methods (i.e., scenario-based design) formalizes activities that would otherwise result in by-products yielding less value. Lastly, externalizing conflicts from people or organizations to artifacts shifts the conflict from a subjective to an objective, more manageable perspective.

References

Badke-Schaub P, Goldschmidt G, Meijer M (2010) How does cognitive conflict in design teams support the development of creative ideas? Creat Innov Manag 19(2):119–133

Brown T (2008) Design thinking. Harv Bus Rev 86:84–92

Carleton T, Leifer L (2009) Stanford's ME310 course as an evolution of engineering design. In: Proceedings of the 19th CIRP design conference
Ciriello RF, Aschoff FR, Dolata M, Richter A (2014) Communicating ideas purposefully-toward a design theory of innovation artifacts. In: European conference on information systems (ECIS), Tel Aviv, Israel, 9–11 June 2014
Dolata M, Schwabe G (2014) Call for action: designing for harmony in creative teams. In: International conference on design science research in information systems. Springer, Cham, pp 273–288
Dolata M, Schwabe G (2016) Design thinking in IS research projects. In: Brenner W, Uebernickel F (eds) Design thinking for innovation. Springer, Cham, pp 67–83
Dolata M, Schwabe G (2017a) Tuning in to more interactivity–learning from IT support for advisory service encounters. i-com. J Interact Media 16(1):23–34
Dolata M, Schwabe G (2017b) Paper practices in institutional talk: how financial advisors impress their clients. Comput Support Coop Work 26(4):769–805
Dolata M, Kilic M, Schwabe G (2019) When a computer speaks institutional talk: exploring challenges and potentials of virtual assistants in face-to-face advisory services. In: Proceedings of the 52nd Hawaii international conference on system sciences
Dolata M, Uebernickel F, Schwabe G (2017) The power of words: towards a methodology for progress monitoring in design thinking projects. In: 13th international conference on Wirtschaftsinformatik 2017, St. Gallen, Switzerland
Edelman JA, Leifer L, Banerjee B et al (2009) Hidden in plain sight: affordances of shared models in team based design. Proceedings of ICED 09, Palo Alto, USA
Gibson JJ (1979a) The ecological approach to visual perception. Houghton Mifflin, Boston
Gibson JJ (1979b) The theory of affordance. In: The ecological approach to visual perception: classic edition. Psychology Press, Hove
Hehn J, Mendez D, Uebernickel F, Brenner W, Broy M (2020) On integrating design thinking for human-centered requirements engineering. IEEE Softw 37(2):25–31
Hevner AR, March ST, Park J, Ram S (2004) Design science in information systems research. MIS Q 28(1):75–105
Johansson-Sköldberg U, Woodilla J, Çetinkaya M (2013) Design thinking: past, present and possible futures. Creat Innov Manag 22:121–146
Karanasios S, Allen D (2018) Activity theory in information systems research. Inf Syst J 28(3): 39–441
Kilic M, Dolata M, Schwabe G (2017) Why do you ask all those questions? Supporting client profiling in financial service encounters. In: Proceedings of the 50th Hawaii international conference on system sciences
Leifer LJ, Steinert M (2011) Dancing with ambiguity: causality behavior, design thinking, and triple-loop-learning. Inf Knowl Syst Manag 10(1–4):151–173
Leifer LJ, Steinert M (2014) Dancing with ambiguity: causality behavior, design thinking, and triple-loop-learning. In: Gassmann O, Schweitzer F (eds) Management of the fuzzy front end of innovation. Springer, Cham, pp 141–158
Leontyev AN (2009) In: Cole M (ed) The development of mind: selected works. Marxists Internet Archive, Pacifica
Lindberg T, Meinel C, Wagner R (2011) Design thinking: a fruitful concept for IT development? In: Meinel C, Leifer L, Plattner H (eds) Design thinking. Springer, Berlin, pp 3–18
Lindberg T, Köppen E, Rauth I, Meinel C (2012) On the perception, adoption and implementation of design thinking in the IT industry. In: Plattner H, Meinel C, Leifer L (eds) Design thinking research. Springer, Berlin, pp 229–240
Luebbe A, Weske M (2012) Determining the effect of tangible business process modeling. In: Plattner H, Meinel C, Leifer L (eds) Design thinking research. Springer, Berlin, p 241
Nicolini D, Mengis J, Swan J (2012) Understanding the role of objects in cross-disciplinary collaboration. Organ Sci 23(3):612–629

Nunamaker JF Jr, Briggs RO, Derrick DC, Schwabe G (2015) The last research mile: achieving both rigor and relevance in information systems research. J Manag Inf Syst 32(3):10–47
Ojasalo K, Koskelo M, Nousiainen AK (2015) Foresight and service design boosting dynamic capabilities in service innovation. In: Agarwal R, Selen W, Roos G, Green R (eds) The handbook of service innovation. Springer, London, pp 193–212
Orlikowski WJ (2010) The sociomateriality of organisational life: considering technology in management research. Adm Sci Q 34(1):125–141
Peffers K, Tuunanen T, Rothenberger MA, Chatterjee S (2007) A design science research methodology for information systems research. J Manag Inf Syst 24(3):45–77
Poole MS, Van de Ven AH (1989) Using paradox to build management and organization theories. Acad Manage Rev 14(4):562–578
Rosson MB, Carroll JM (2009) Scenario based design. Human-Computer Interaction, Boca Raton, pp 145–162
Sambamurthy V, Zmud RW (2012) Guiding the digital transformation of organizations. Legerity Digital Press, Tallahassee
Schwabe G, Nussbaumer P (2009) Why IT is not being used for financial advisory. In: 17th European conference on information systems (ECIS 2009), Verona, Italy
Smith W, Erez M, Jarvenpaa S, Lewis M, Tracey P (2017) Adding complexity to theories of paradox, tensions, and dualities of innovation and change: introduction to organization studies special issue on paradox, tensions, and dualities of innovation and change. Organ Stud 38(3–4): 303–317
Uebernickel F, Brenner W (2016) Design thinking. In: Hoffmann CP, Lennerts S, Schmitz C, Stölzle W, Uebernickel F (eds) Business innovation: Das St. Galler Modell. Springer, Fachmedien Wiesbaden, Wiesbaden, pp 243–265
Vetterli C, Uebernickel F, Brenner W (2012) Initialzündung durch Embedded Design Thinking—Ein Fallbeispiel aus der Finanzindustrie und wie dadurch ein Wandel in der Innnovationskultur einer IT-Abteilung eingeleitet wurde. Zeitschrift für Unternehmensentwicklung und Change Management 2:22–31

Platform Design with Design Thinking and Scrum: An Experience Report from Deutsche Bundesbank

Michael Jakob and Jennifer Hehn

Introduction

Digitalization is transforming the financial sector. Technological developments like cloud-based solutions and microservices demand for a reconsideration of the status quo, also in the stability-oriented sector of central banking. The Deutsche Bundesbank is the independent central bank of Germany with the main objective to secure price stability in the euro area. In addition, the Bundesbank performs other key tasks such as national supervision of credit institutions, cash management, payment systems and financial stability. Moreover, the Bundesbank manages Germany's foreign reserves, acts as the government's fiscal agent, and carries out important statistical tasks (Deutsche Bundesbank 2020).

In 2018, Bundesbank started a broad digital transformation program to leverage the new possibilities of technological advancements and to foster innovation and customer-centricity. Therefore, Bundesbank decided to redesign its ExtraNet, the largest interaction platform between Bundesbank and its more than 100,000 customers, including insurances, public services, monetary finance institutions, and other companies. The goal of the new platform was to provide (new) user-oriented services and a user-friendly interface based on a modern IT infrastructure.

Initiated by a then new member of the executive board, the Bundesbank decided to utilize the explorative approach of Design Thinking to rethink the purpose of the platform, gain better insight into customer needs, and to generate new ideas for new

M. Jakob
Deutsche Bundesbank, Frankfurt, Germany
e-mail: michael.jakob@bundesbank.de

J. Hehn (✉)
Institute for Digital Technology Management, Bern University of Applied Sciences, Bern, Switzerland
e-mail: jennifer.hehn@bfh.ch

J. Hehn et al. (eds.), *Design Thinking for Software Engineering*, Progress in IS,
https://doi.org/10.1007/978-3-030-90594-1_10

ways of user-centric interaction. To that point in time, Design Thinking as well as collaboration in a multidisciplinary project team spanning departments across the entire organization was new to Bundesbank.

In this contribution, we summarize our research protocol and provide a detailed overview of how Design Thinking was used in combination with Scrum on a process, toolbox, and mindset level. Our report expresses the authors' personal impressions and opinion and does not necessarily reflect the view of Deutsche Bundesbank.

Project Context and Description

The project to design a new Extranet was initiated based on the need to renew the existing eBusiness-Platform ExtraNet of Deutsche Bundesbank. The current ExtraNet has been running reliably for more than 18 years while providing shared services (e.g., file transfer, user management) and interactive applications (e.g., bidding systems, reporting platform of banking supervision and statistics). The ExtraNet is a high-performance data exchange platform used by internal applications of different business areas. It is provided by the internal IT department as an on-premise solution including the platform itself, the infrastructure, the application services, and operational tasks. The reports exchanged between external customers and the Bundesbank range from six to ten million per year. The renewal of the ExtraNet became necessary from both a business and technical point of view.

The project manager and sponsor decided to apply Design Thinking to better understand the problem domain before drawing conclusions too quickly for a possible solution and IT architecture. Keeping in mind that the data exchange platform started small and had developed into a large, business-critical Bundesbank solution, including heterogeneous external customer groups and different internal business areas, the conditions and technological possibilities for the platform had completely changed over time.

Until now, the project is divided into three phases: (1) *Exploration*, (2) *IT-Prototyping*, and (3) *Final Prototype*. Phase (4) *Implementation* is currently in planning. Table 1 provides an overview of each project phase with its respective goal, activities, and deliverables.

In the following, we present and discuss each project phase along the structure of (1) objective (2) prerequisites, (3) activities, (4) roles, (5) outcome, and (6) reflection for each phase.

Phase 1 "Exploration"

Objective The goal of the exploration phase was to understand the problem and needs of the platform users and to create a new clear product vision.

Table 1 Overview of each project phase with goal, activities, and deliverables

Project phase/activities/deliverables	FTE[a]
Phase 1: Exploration (09/2018–12/2018) *Goal: Understand customer needs and define a clear product vision*	10
• 16 qualitative interviews with external users • 5 qualitative interviews with internal users • 37 insights, 5 personae, 3 visual frameworks • 10 internal technical insight sessions • 10 internal business expert sessions • Benchmarking and technical market analysis • 7 opportunity areas for new solution development • 17 brainstorming sessions and 402 ideas • 12 low-resolution prototypes • 3 aggregated prototypes • 20 test sessions with customers • Technology screening for IT architecture • 2 management presentations ("walk-ins")	
Phase 2: IT-prototyping (01/2019 —09/2019) *Goal: Develop a proof-of-concept and proof of architecture*	30
• 4 prototypes (1 failed) derived from Design Thinking • Experience exchange with 3 companies • Development of three IT-prototypes • Implementation of infrastructure platform • Test sessions with 31 external customers • Test sessions with 55 internal business experts • Technology screening with an external agency	
Phase 3: Final prototype (04/2020 —03/2021) *Goal: Build and test a final IT-prototype integrating all core functionalities*	20
• Market analysis with external consultancy • 12 sprints applying scrum and 382 story points • 31 epics and approximately 700 user stories • 1 application prototype based on 100 user stories • 4 IT infrastructures (e.g. Public Cloud, Private Cloud) built and used by above mentioned application • Integration of 3 backend applications • Testing with 100 internal business experts • Testing and feedback with 20 managers • Build-up product backlog (700 user stories) • 5 lessons learned workshops	

[a]Full-time equivalent (FTE)

Prerequisites Before kicking off the project, a design challenge was defined, a multidisciplinary team was formed, a project war room was booked, management support was secured, and Design Thinking coaches were contracted. The design challenge inhabited an exploratory character to provide direction, yet to leave room for unpredicted discoveries ("How might we redesign the eBusiness-Platform of Deutsche Bundesbank?"). The core project team was composed by involved business areas, core central banking functions (banking supervision, cash, statistics), and IT-experts from development, infrastructure, and project management. A so-called extended team was formed to provide fast and easy access to a pool of experts and management. A large office space was booked to enable collocation and collaboration as the project members worked at different Bundesbank locations. IT-Management supported all activities and sponsored the external recruitment of Design Thinking coaches to lead through the innovation process.

Activities The activities followed the Design Thinking process of problem definition, need-finding, synthesis, ideation, prototyping, and testing. The team conducted contextual interviews with (potential) internal and external users of the E-Business platform. The latter included employees of banks, insurances, public services, and companies who file reports and messages to Bundesbank. The main purpose of the interviews was to identify their needs regarding the existing ExtraNet in particular, and the opinion concerning the digitalization ability of Bundesbank in general. The project team collected improvement proposals and insights, e.g., "We want Bundesbank to provide transparent processes and one face to the customer." The different quotes of the interviewees were clustered, transferred into insights and aggregated to opportunity areas (e.g., "get rid of paper"). The team designed five personae and customer journeys to provide a better visual impression of the most interesting findings and potential value propositions. While the personae inhabited the thoughts and feelings of different user types, the customer journey showed the relevant usage steps from registration to customer support and log out. To develop new ideas and features the extended team was invited for several brainstorming sessions based on predefined ideation questions. Another aspect was to gain internal technical and business knowledge and integrate colleagues in the new Design Thinking approach. Based on these ideas, the project team created paper-based prototypes, which were tested and refined iteratively according to the received feedback. Some of the current services, e.g., the user help desk and the reliability of the ExtraNet were highly appreciated. Some wishes for the future solution were mentioned by many customers, e.g., easy-to-use security features and to be able to use one portal access for all use cases (one face to the customer). Eventually, the most promising elements of the prototypes were aggregated, while others had failed and were discarded. The results were presented to Bundesbank management in a "walk-in" session at the end of phase 1.

Roles During the exploration phase, each team member was involved in all Design Thinking activities, regardless of their expertise. The goal was to gain the same level of empathy for the affected customers and the problem statement. The project manager reported the progress and the results to the organization. The Design Thinking coaches were externally hired and supported the team along each step of the process as there was no prior experience with Design Thinking at Bundesbank.

Outcome The deliverables of the Design Thinking process were three non-technical prototypes called "Octopus," "User Manager," and "Transparent Process" (see Fig. 1). These prototypes represented the key specifications and needs discovered during the need finding activities. The "Octopus"—having many arms to contact others—was used to communicate the technological advantages of a

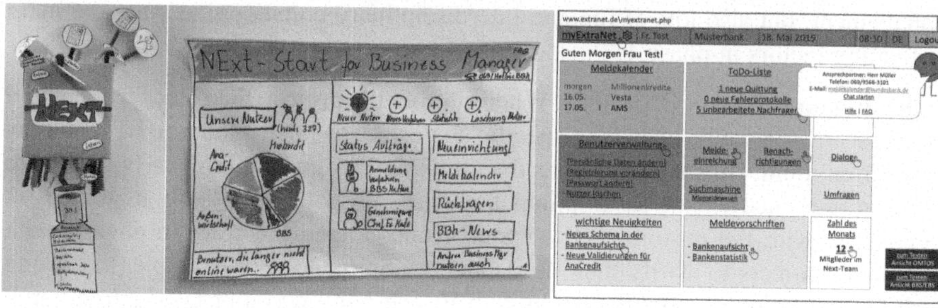

Fig. 1 Overview of three non-technical prototypes (from left to right: "Octopus," "User Manager," and "Transparent Process")

modern IT platform. The team used paper wireframes and, in the second step PowerPoint with interactive functions to specify the ideas of a "User Manager" as a self-service solution and the "Transparent Process" as a user-friendly and seamless platform experience.

Reflection on Methodology During the first phase, the Design Thinking approach was completely new to Bundesbank. The team was able to use an office space exclusively for the workshops and results. This was an important precondition for being successful as it saved a lot of time when keeping results on boards and walls. It turned out to be a safe haven for trustful communication within the team and with colleagues. It was the ideal working area to work out our Design Thinking tasks unhindered from the organizational line to which each team member belonged. Another success factor was the hiring of external Design Thinking coaches in time, who fitted to the team and brought the needed practical experience in applying Design Thinking. The need identification techniques (e.g., interviews, ideation, prototyping) were helpful and allowed a structured way of doing the job. The prototypes, as a result, reflected the customer requirements. As they were tested several times, the team perceived that there was little to no risk in omitting customer wishes by substituting needs through predefined internal objectives.

Reflection on Teamwork Design Thinking activities felt exhausting to many team members, as they were not only new but also performed in addition to day-to-day tasks. Nevertheless, the team identified quickly with the mission and was willing to put in the extra mile. One reason was the aha-momentum when interviewing customers face-to-face and listening to their needs. This activity felt convincing and inspiring. The team undertook a commitment to improving things not only for the IT department or for internal business areas but also for all customers of the platform. The interdisciplinary work for such a long period of time was not common at Bundesbank, but the team perceived it as urgently needed to understand different perspectives. The customer's voice provided common ground for the team, and, quickly, all team members grew together and reached the performing phase. Members stated that openness, responsibility for the result, experiencing customers face-to-face, encouraging creative methods, high expectations and motivation from management, but also lots of fun and interdisciplinary collaboration were the main motivating factors in phase 1.

Reflection on Business Objective When starting the project, the IT department and business areas already had their specific expectations of a modernized and new platform in mind. However, the objective of using Design Thinking early on was to question previously made assumptions and come up with completely new ideas in co-creation with the customer. The sponsor encouraged the team to engage with external customers to clarify their needs and requirements. For example, the strong wish for one point of communication with Bundesbank was translated into a portal with full transparent processes and a status overview at any time. The high level of interest in Design Thinking in phase 1 resulted in multiple (intermediate) presentions to inform stakeholders and interested colleagues about the activities and intermediate project artifacts, including European colleagues and the managing board. An important artifact were distinctive quotes of customers, which superseded any further discussions. Another convincing method was prototyping and testing. The prototypes acted as communication starters and sparked new discussion topics and feedback. Not only did they provide a clear impression of the new platform and the needed services, but they also revealed attitudes, arguments, and key aspects, especially of management. When testing prototypes, we gained feedback for our ideas and strengthened our network at the same time. However, we the team also faced rising expectations for integrating the elicited requirements.

Phase 2 "IT-Prototyping"

Objective The main goal of the second phase was to provide the proof-of-concept and architecture for the created prototypes. In comparison to the Design Thinking phase, a higher level of fidelity concerning the IT-prototypes and their functionalities was the goal. Finally, the results, including the feedback of the customers and the IT-prototypes, should lead to one overall solution proposal for the following phase. A benchmark with external support and exchange was included to ensure a modern architecture.

Prerequisites Three validated non-technical prototypes in the form of paper and one clickable PowerPoint presentation as a high-level system vision were the core input for phase two. To continue with the open working spirit and user-centered mindset of the first project phase, nearly the complete Design Thinking team continued to work on the project. We added internal developers, business-related experts, an IT architect, and a security expert to focus on evaluating the feasibility of the ideas.

Activities The iterative approach of prototyping, testing, and refinement was the guiding principle of the second project phase. Three interdisciplinary teams were formed around one prototype from phase one. Each so-called "prototyping team" clarified the requirements for their assigned prototype and started to develop an IT-prototype. A fourth subgroup consisted of IT engineering colleagues who provided the needed development tools and infrastructure platform. The existing Design Thinking prototypes were explained to the new team members. Customers were invited to verify former ideas and add new insights. The development of the so-called IT-prototypes should consist of developed programs running on middleware and infrastructure. After the basic IT tools and components were established, the prototyping teams immediately started developing software based on validated and concretized requirements. Whenever a particular IT prototype of a subgroup was executable, customers were invited to test them directly. The test sessions and results were recorded along results, new ideas, improvement proposals, and fails. The core IT services were checked and discussed with Gartner architects and security experts. State-of-the-art services and software components were identified, among them, for example, Identity and Access Management (IAM), Application Programming Interface (API)-techniques and microservices. During this phase, the non-functional and functional requirements were specified in more detail. Furthermore, exchange with companies with comparable customers, and business objectives were undertaken and delivered insights into the market and technical approaches (technical benchmarking).

Roles The organizational chart included four sub-teams staffed with team members from the previous phase and new experts: one team provided the IT infrastructure and development tools and three teams were to specify and develop three IT-prototypes. An IT-enterprise architect and an IT security expert were onboarded. The IT project manager continued to stay on the project. In this phase, no specific coach was involved.

Outcome The feasibility study with implemented IT-prototypes was successfully realized. The proof-of-concept was provided by implementing an IT platform and three IT-prototypes running on it: (1) a technical API for one specific business area ("Octopus" from phase one), (2) a transparent process with status information and a landing page ("Transparent Process"), and (3) a user manager self-service including 2-factor-authentification (2FA) for external customers ("User Manager"). The idea of "one face to the customer" inspired the team to offer shared portal services. The front-end was programmed using Angular. A middleware solution was used for communication services. Each prototype was tested after being developed successfully in test sessions with internal and external customers.

Reflection on Methodology This phase was guided by no specific method. Being the outcome of phase one, the Design Thinking prototypes and the principle of prototyping—testing—refining was applied iteratively and successfully in each of the subgroups. It worked well and turned out to be the common ground, also for onboarding new team members. However, the mindset that had been internalized by the team members of the previous phase could not be transferred completely to all new team members as too much implicit knowledge was missing. Accordingly, the former Design Thinking team members did a lot of explaining and reasoning at the beginning. However, the new team members were convinced when testing started, and customers were invited to provide feedback. The development was performed in the background by an internal IT unit. To ensure IT-alignment developers and business experts met regularly to define requirements ready for implementation. There was no Scrum or Kanban approach. The results of the subgroups were shared in a so-called "Design Thinking Board," an overall board focusing on user needs and requirements.

Reflection on Teamwork The main challenge of the Design Thinking team was to transfer the principles, needs, and ideas to the new team members, e.g., to developers never having applied Design Thinking or a comparable user-centric approach before. At first, they were skeptical and focused on development methods. The close interdisciplinary collaboration in the subgroups with at least two members from the Design Thinking phase ensured keeping the open and user-centric spirit alive. Later, the interviews and test sessions with real customers were convincing for all team members and reconfirmed that the prototypes fulfilled the customer needs.

Reflection on Business Objective From the business perspective, the most important aspect of phase two was to validate the feasibility and viability of the developed ideas. This meant that the prototypes from phase 1 could be developed, transferred, and implemented with Bundesbank's IT resources and techniques. The creation of IT-prototypes validated the Design Thinking prototypes from phase one, which was an important milestone in removing concerns about the effectiveness of the new methodology of Design Thinking. Again, there was the notion that the customers who supported cooperatively in test sessions expected the project to deliver results quickly.

Phase 3 "Final Prototype"

Objective After having confirmed the feasibility of the prototypes, market analysis and benchmarking should make sure that we did not overlook any new technical or business trends. The essential output of this phase was to build and test a final prototype integrating all core functionalities based on the previously identified needs and requirements. The final prototype should include an IT architecture and technical connections to real backend applications. If successfully evaluated, the result should be the proposal for the productive IT platform to be implemented and prepared for a productive Go-Live.

Prerequisites The findings, results, and customer feedback from the previous phases were transferred into the third phase. The project team changed again. Five individuals of the initial Design Thinking team remained in the project. Internal developers left the team. Scrum was selected as an agile framework, although this—again—was a new method for the team. The overall vision was to deliver business and customer benefits by working out one functioning IT-prototype that should represent the idea of one-face-to-the-customer with easy-to-use and fully transparent services based on a technical state-of-the-art cloud-based solution.

Activities As the remaining Design Thinking team members were experienced with the mode of prototyping, testing, and customer-centricity, each one took over the role of the Product Owner, forming a so-called Product Owner Team (POT). Being a precondition for the later steps, the project team initiated a market analysis and spent 4 months in close cooperation with an external consultancy. Starting with a detailed explanation of business requirements, the analysis should also evaluate common IT provider concepts (public, hybrid, private cloud), IT-service models (IaaS, PaaS, SaaS) and state-of-the-art software products. The task included a solution recommendation describing which provider model and software products should be applied when designing the final IT-prototype. In the meantime, the POT started to define the product backlog. Many business requirements could be extracted from the former IT-prototypes and the legacy platform. A scrum master coached the POT. In the second half of this phase, the development of the final prototype took place, which included implementing the IT infrastructure and programming tasks. Many cloud techniques (e.g., DevOps, microservices) were used. External developers were hired. They were also accustomed to Scrum. In 12 sprints, the work was completed. The POT defined the complete business process and user stories. While the code was developed iteratively, the different IT infrastructure components were also built up in sprints and tested. They were based on private and public cloud platforms. After the last sprint, several internal customers were invited to test the final prototype and to provide feedback. At the end of this phase, several lessons learned workshops were conducted to preserve the multitude of new insights regarding content created and methods used.

Roles The POT was again set up as an interdisciplinary team and responsible for collecting and prioritizing user stories as well as communicating with stakeholders. A scrum master was hired to implement the new agile method and role. The product backlog was the first artifact created. Apart from experienced external developers, four internal colleagues reinforced the development team. A project manager, the project sponsor, an IT security expert, and an IT architect completed the team.

Outcome The market analysis was an important conceptual output. The final prototype was the physical result. It covered the specification of the designated IT solution, the core functionalities, and IT architecture. The new cloud techniques, e.g., microservices, API, DevOps, turned out to work well and fit the requirements. From an IT point of view, it was important to evaluate the provider and service models of private and public cloud and SaaS. A convincing result was the portability of the programmed application code, which was executable on four different (private and public) cloud platforms. The business case was validated by internal business experts. The prototype was a suitable way of communicating and experiencing the future product vision. Scrum artifacts like the Product Backlog were created and developed further.

Reflection on Methodology Scrum was perceived as the right method to ensure an agile way of working. It was new to the internal team members. Although it took two sprints to get used to, the POT, Scrum master, and developers started to speak a common language based on the Scrum terminology (e.g., review, daily, retrospective). A survey at the end revealed the good mood of the team and the belief that openness, trust, and transparency are success factors also to apply scrum successfully. The best practice of phase two of "prototyping-testing-refinement" made sure that the needs and requirements of customers were integrated continuously.

Reflection on Teamwork In a short period of time, again, the team had to learn, adopt, and apply a new method and work together with external experts who needed to understand the business requirements. This experience and the visible project progress strengthened the trust in agile and innovative methods such as Design Thinking and Scrum. The mixture of the team, consisting of business and IT, internal and external, experienced, and young personnel seemed to be a challenge but it turned out to be an advantage and essential success factor. A high availability of the involved staff (>70%) was an important precondition when applying Scrum and not always easy to be enforced.

Reflection on Business Objective The final prototype was successfully developed and implemented, including the provisioning of different cloud models. The business process ran smoothly. New techniques and tools were provided, all of which were new to the organization and therefore perceived risky. The team demonstrated that the development and implementation, which took at least 6 months, including around a hundred user stories and four infrastructure platforms (Public Cloud, Private Cloud Bundesbank), was feasible. The connection to the backend applications showed interoperability of the new cloud technique and the old on-premise world. From a customer point of view, the portal represented the need "one face to the customer" and "transparent processes" including different business services in one integrated IT-prototype. The business process ran successfully from registering users up to status information returned by backend applications. Feedback given by internal business experts reconfirmed usability and operability.

Outlook for Phase 4 "Go-Live"

The final prototype had turned out to be desirable from a user point of view, feasible from an IT perspective, and viable from a business standpoint. In addition, the advantages of using Design Thinking for problem and solution exploration and Scrum as an agile implementation approach were confirmed by the project team. In the final project phase, Go-Live, the team plans to realize the system vision of a *user-centric, state-of-the-art E-Business-Platform which delivers reliable, secure, and transparent services to external customers and internal business experts*. The main technical task will be the design and implementation of the production platform. The project team, again, will be an interdisciplinary one working with Scrum and the validated approach of "prototyping-testing-refinement." After team building, one of the first activities will be the setup of a cloud-based platform and the development of the minimum viable product (MVP). This will enable the team to keep testing new features, especially with external customers. In addition, the integration of internal applications will be performed. This step will close the gap to the backend processing and deliver a technical integration of the backend and the outside world. Concerning the organizational structure, the digitization of existing paper-bound processes and the new level of automation (e.g., through chatbot-functionalities and application-to-application integration) will deliver significant improvements compared to today. Additionally, the relevance of customer satisfaction and the perception of Bundesbank as an innovative and reliable service partner cannot be overestimated. Customer acceptance will be checked continuously by analyzing usage patterns, surveys, interviews, and so on. The migration scenario will be a large work package to be agreed upon with each business area. Whatever it takes, there is the strong confidence that the MVP is close to the user expectations as expressed during Design Thinking, while the agile approach will allow for any needed adaption further down the line.

Key Learnings

Reflecting on the project so far, we want to summarize and highlight the following key learnings.

Consistency of the Team Along the Project Is Needed to Keep the User-Centric Mindset Alive Parts of the project team changed from phase to phase. The risk of losing Design Thinking members was identified early on. We made sure to keep 50% of the initial team on the project throughout all phases. This not only provided a continuation of the acquired knowledge about customer needs and requirements but also manifested a team mindset characterized by openness, trust, creativity, and radical collaboration. This spirit motivated new team members and convinced them to explore unknown methods.

Cross-Functional Work and Different Perspectives Are Key to Creating Innovative Solutions From a business objective, the bundle of new methods and time spent for collaboration across different business areas seemed to be challenging at first. However, integrating customers and all relevant units continuously provided ideas that had not been considered before. In addition, the constant collaboration helped project members to identify themselves with the product in a way we had not experienced before.

The Continuous Approach of "Prototyping-Testing-Refinement" Provides a Core Connection Between Methods Across all phases, most of the time was invested in building prototypes, starting from (non-technical) low-resolution mockups to the final (technical) IT-prototype. The iterative approach of "prototyping-testing-refinement" revealed itself to be the common thread among all project phases linking Design Thinking and Scrum. In doing so, the team ensured a continuous level of creativity, ongoing interaction with users, and concurrent development of new ideas and features. We feel confident now that the future system—when following the same rules and principles—will be a success.

Design Thinking and Scrum Motivate to Go the Extra Mile Although the different working methods seemed exhausting first, they proved to be great motivational factors to keep the team spirit high. Considering the novelty of methods used, stakeholder management, and actual project work, we estimate the net time for getting the job done at around 18 months, starting with Design Thinking until finishing the final IT-prototype. The rest of the time was spent on internal processes (e.g., securing approvals and dealing with procurement).

Non-Functional and System Requirements Are a Challenge What remains challenging to us is the identification and specification of non-functional requirements in the context of Design Thinking and Scrum. We consider this task not just difficult in general but especially in the context of our agile and user-focused style of working. For example, we felt it quite challenging to combine our user stories with architectural or system requirements. This could be a promising topic for further exploration.

Reference

Deutsche Bundesbank (2020) Central Bank of the Federal Republic of Germany. Available at: https://www.bundesbank.de/en/tasks/central-bank-of-the-federal-republic-of-germany-626946. Accessed 2 May 2020

Design Thinking in a Large Manufacturing Organization: Designing a Smart Support System for the Shop Floor

Markus Durstewitz and Thomas Abrell

Introduction: Design Thinking in Manufacturing

Industry in general and large manufacturing companies in particular are focusing on operational excellence (Issar and Navon 2016), namely on continuous improvement of process efficiency, aiming for maximizing the output while minimizing the applied effort in terms of resources, such as energy and manpower. Workers are there to work. And they should do good work. More precisely, it is expected that they execute their job with excellence exactly as described and that they deliver within the predefined time frame. High quality outcomes are expected, and safety considerations leave no room for interpretations. Continuous improvement schemes are aiming for process and lead time optimization, avoiding any safety issue while minimizing the probability and cost of time delays and non-quality.

Unfortunately, real life working conditions are very often not the same as the ones assumed and predefined by the manufacturing engineer, particularly in the context of complex products and systems, where manufacturing takes place in comparatively small batches (Hobday 2000). Actual work situations can differ from day to day due to unforeseen events that risk occurring at a certain probability. Thus, non-conformities cannot be fully excluded due to the nature of complex work resulting in limitations when targeting an absolute control of the working environment. Unforeseen events triggered by environmental or human factors lead to

---

Before at Airbus Operations GmbH, involved as project lead in the case described and therefore independent of the author's current affiliation with Volkswagen AG

M. Durstewitz (✉)
Airbus Operations GmbH, Hamburg, Germany
e-mail: markus.durstewitz@airbus.com

T. Abrell (✉)
Volkswagen AG, Wolfsburg, Germany

J. Hehn et al. (eds.), *Design Thinking for Software Engineering*, Progress in IS,
https://doi.org/10.1007/978-3-030-90594-1_11

abnormal situations in job execution, which may result in process disruptions, quality, and/or safety issues (Hollnagel 2009). Then, the question is how to deal with these unforeseen abnormal situations? The experienced worker knows how to overcome unpredicted problems. But unfortunately, the workers themselves are not systematically involved in the design of the process and only trained to be focused on executing the job as prescribed.

Participatory design (Grudin 1993) formally integrated user experience in the design of human–machine interactions (Durstewitz 1994). Today, Design Thinking[1] (Brown 2009) represents a holistic human-centered design approach. It helps to leverage the practical experience of workers that often remains implicit hidden knowledge (von Hippel 1994) and as such it is not or is only used by chance for the design of the workflow and work environment.

In the design of software-intensive digital solutions, the designer has to deal with interaction modes at cognitive level that are more difficult to uncover and therefore tend to remain more often tacit and hidden (Boy 2011). In addition, software-intensive solutions are used in general to support, control, and optimize workflows and manufacturing execution. Therefore, they have to represent the real processes and map them onto their inherent respective data models and algorithms as close as possible to reality. Thus, the better we know what the worker thinks and which rationale the worker applies when assessing a given situation, the more effective and affordable will be the solution we design for this situation.

Design Thinking is radically user centered. In consequence, the workers' perspective becomes a fundamental and essential design input. And Design Thinking embraces a holistic empathetic approach encompassing all different aspects of the user explicitly including the cognitive and emotional side, thinking and feeling. For this, the workers must be directly involved in the design process itself with a clear focus on their own practical work experience thus ensuring that workers are working on the right things (Abrell and Durstewitz 2016), i.e., on the things that matter to get the job done at any given point in time. Using Design Thinking reduces the risk of late and then very costly design changes through consequent user involvement along the complete development process starting with the project framing and early prototyping. Compared to other agile development approaches, Design Thinking radically emphasizes the importance of user insights, continuous user participation, and direct user feedback for each cycle, from the (re)definition of the problem to be solved to the design and validation of the respective most acceptable and therefore most valuable solution. For this, Design Thinking relies on open collaboration and co-creation of multifunctional teams (Leifer and Steinert 2011).

With the here presented case we tried to go further applying Design Thinking to the very heavy industrial organization of manufacturing, the design of a new final assembly line (FAL) for a very complex large product.

[1]For an introduction to Design Thinking, see Brenner et al. (2016)

Description: Our Approach to Design Thinking in Three Phases

In our specific case, we followed the basic process of the Design Thinking approach (Brenner et al. 2016) with a special emphasis on end user engagement through explorative need finding and co-creation, and therefore focusing on prioritization of the jobs to be done, i.e., "Doing the right things" before focussing on optimization, i.e., "Doing things right." In addition, from a project management perspective we needed structure to be in synchronization and alignment with the predefined overall milestone plan of the new FAL global lighthouse project. Thus, we organized our Design Thinking approach in three subsequent independent phases: Phase1 "Need Finding," Phase 2 "Prototyping & Testing," and Phase 3 "Implementing."

Each phase itself consisted of a divergence and convergence part (inspired by the double diamond scheme (Design Council 2019)) with several agile iteration cycles, and with respect to Design Thinking always with a clear focus on field activities and deliverables applicable in the local practice on site, aiming for gathering a maximum of user insights and direct user feedback. Finally, each phase was expected to provide its respective deliverables as input to the global development project.

Phase 1 "Need Finding"

The major objective of Phase 1 "Need finding" was the framing of the problem and the definition of one or several concrete problem statements reflecting the essential, previously hidden and tacit, user needs. We wanted to gain understanding of the organizational setting (role), mindset (motivation), and skillset (capabilities) of different users of the final assembly line in their respective working environment and under the given working conditions. Who are the users? How and where do they work? Which jobs do they execute (user segments) that we need to include in our Design Thinking approach to cover all individual perspectives and a holistic view of the jobs that need to be done (Ulwick 2016)?

Getting access to end users, in our case ultimately the shop floor workers, and directly engaging them in need finding and co-creation is the key factor of success and key differentiator of Design Thinking compared to other, namely engineering-driven development approaches. Therefore, first we needed to build trust and a good relationship with all key stakeholders and in particular with the end users. This step might sometimes be very time consuming. But it is utmost critical for the success of any Design Thinking project. In our case, it was also a unique opportunity for promoting Design Thinking while contributing to the design of the new FAL. As a first contact and engagement measure, we convinced the project leader to conduct a co-creation workshop using Lego Serious Play (LEGO Group 2010) in order to clear the project mission, to understand the local environment, working conditions and specific project needs with respect to user engagement. As a result of our effort, our approach was adopted and we could initiate a dedicated work stream called "FAL 4 Workers." As such the Design Thinking team was formally integrated into the global project structure participating in regular project meetings and progress reviews. In addition, this enabled us to get introduced by the team to the end users, the shop floor workers, becoming an intrinsic part and members of a multifunctional Design Thinking team.

Need Finding itself was conducted during 3 months with only 8 weeks for in-depth field observations in one of the existing final assembly lines. This is a relatively short period of time but with a high intensity enough for encompassing most work situations including all shifts, day and night. Doing so, we gained insights on the users' work, working conditions, the shop floor organization, while accessing both explicit knowledge such as organization charts, process documents, workflows in IT systems, as well as tacit knowledge of the workers. In order to understand the complex environment from a user perspective, we observed the users, shop floor workers, and first line managers in their actual work context, we shadowed them, and finally conducted contextual interviews (Stickdorn and Schneider 2010). During all activities, we focused on reaching a general understanding of the entire manufacturing and assembly process from a user perspective and gaining specific insights on crucial parts of the process. Parallel to the observations and field work, we conducted an ongoing analysis through affinity-mapping (IDEO 2015) and extensive discussions with users, and therewith triangulating insights and identifying the most critical themes for further investigation. Then, we distilled the insights to an opportunity map and associated user journeys (Stickdorn and Schneider 2010). We created a detailed documentation of each potential opportunity area within these user journeys. Based on the map and user journeys, we conducted a co-creation workshop explicitly including the actual end users and works council representatives to prioritize the most relevant opportunity areas. All participants agreed that "Information Exchange," the usage of and access to data and information, represented the most critical area. The project team identified two major use cases to be both of high priority and possible to implement in the project timeframe: logistics (material flow and availability) and exceptions (non-conformity) management.

Phase 2 "Prototyping & Testing"

Compared to other "classical" prototyping approaches, Design Thinking is radically user centered. The users are always kept in the loop starting with co-creating experiments to validate assumptions, to pivot and subsequently mature solutions. The major objective of the prototyping phase is the exploration of the possible solution space and the selection of the most appropriate solutions, i.e., the ones matching the problem statement being the right thing to do, and the most promising solutions, i.e., the ones showing highest evidence and effectiveness by creating the biggest impact in solving the problem or getting the job done, always based on the direct feedback of the users. This included the eventual adaptation of the original work environment and working conditions that best respond to the real (otherwise hidden) user needs, i.e., which are most effective, and in return create the highest value for the company, for example, by avoiding cost for non-quality or useless investment for ineffective and ultimately not used solutions. The first action of Phase 2 was to enrich the findings of Phase 1 with deeper knowledge concerning the specificity of the information management actually used by the workers and their respective need for accessing specific information at critical points of the job to be done. This was not limited to digital solutions but explicitly included offline and paper-based solutions in use, for example, individual hand-written notes. In addition, information was gathered about the existing IT systems, the major pain points linked to the use of these systems and not accomplished information needs. The Design Thinking team was now complemented by specialists from a digital design agency with focus on user-centered development of IT solutions, which brought new expert knowledge and beforehand missing digital prototyping capabilities to the team. But it also required a ramp-up time for the new team members to become operational. In the end, the newly formed team was synthesizing the insights from general need finding and the specific research into design drivers. Then, the team identified core problems and ideated for possible solutions. The ideas were prioritized considering all three basic Design Thinking perspectives: user desirability, business viability, and technical feasibility (Brown 2008).

On this base, the team started the prototyping activity, organized in cycles. Each cycle started with user insights, leading toward new prototypes, and closed with a co-creation workshop to collect user feedback, to validate the prototypes, and to initiate the next iteration. Based on the outcome of the respective cycle, the team was able to incrementally increase the maturity of the subsequent prototypes. In our case, in the short time span of 12 weeks, three prototyping cycles were conducted. The first cycle delivered a low-fi paper prototype, ideas, and story boards addressing the use of different devices for the selected use cases (see Fig. 1). This included the use of individual smartwatches or smartphones, a station tablet, an interactive material rack, and an interactive team board. The different devices were all assessed and different user flows prototyped. Then each idea was evaluated by the users concerning its usefulness for their daily work and respective jobs to be done. The interaction with the tangible prototypes helped a lot unlocking the tacit user knowledge and quickly progressing toward the next iteration.

At this stage, the smartwatch was excluded because of occupational safety concerns by the users and at that time too limited functionality. The material rack was discarded because the users did not see any additional value compared to the use of the station tablet, which at the same time offered more flexibility and was easier to implement. Thus, it was decided to focus on the design of a mobile app for use on individual smartphones and on the station tablet, and to implement the interactive team board allowing to collect all individual information and collectively assess the actual work status and exceptions by the shift teams during the daily briefings. It was important to continue involving all other stakeholders to ensure their commitment and acceptance for implementing proven solutions. For this purpose, a demo day was organized on the shop floor to gather feedback for the working final prototypes

Fig. 1 Different devices were evaluated, different user flows prototyped

in close proximity to the real working environment. The outcome of the demo day materialized in a modular information board for each manufacturing station that showed the respective staff at work and their tasks as well as the material status and possible identified exceptions.

Furthermore, we validated the introduction of the mobile app enabling workers to receive relevant information during their shift, but also to report possible issues or to ask for support. It was implemented with a working back-end application, microservice architecture, and matching API's. The final outcome was fully documented and represented as the tangible result and deliverable of the prototyping & testing phase, ready for implementation.

Phase 3 "Implementing"

The major objective of Phase 3 "Implementing" is the integration of a pilot application into the operational environment and its adoption by a launching "customer" by the means of a Minimum Viable Product (MVP, the term is used by the Lean Startup methodology (Ries 2011)). Here, customer means the respective operational application area, in our case the final assembly line. The pilot application approach guarantees fast adaptation rates and a steep learning curve, while limiting the risk of failure to one selected application area. The MVP is limited to the key functionalities necessary to solve the problem, in our case providing information about the job organization and material availability to the worker at the point of use and allowing them to record and report any exceptions and non-conformities (see Fig. 2). For the operational proof of software-intensive products and services it is critical to access real production data, in general provided by the company's enterprise resource planning system. This is often very difficult requiring the integration both in the

Fig. 2 Input for implementation of pilot applications as MVP

real working environment and in the existing (sometimes outdated and closed legacy) information system infrastructure and architecture.

For the MVP, it is recommended to select a minimum number of core functionalities necessary to solve the identified problems. In general, this starts with one functionality and successively new functionality is added to a point that finally the application creates the impact expected by the end users while remaining easy to operate at a level that is accepted by them. In our case, the first functional pilot application just delivered clear work instructions for the job to be done and enabled the worker to record and report any exception by simply taking a picture of the non-conformity, adding personal comments and the exact location by just some clicks, and all this without leaving the actual work position. In the end, Phase 3 could not be completed by the Design Thinking project team as initially planned. Instead, it was handed over to the Manufacturing Execution System (MES) project that had been launched in parallel at a higher level to renew the overall MES. This larger project was prime and part of the company's global Digital Transformation initiative at group level.

Reflection: Critical Success Factors and Global Analysis

Even though we know that some findings are causally linked to the organization and the specific context of our case, we believe it may serve as a general blueprint for similar undertakings applying Design Thinking in industrial settings.

Let us first look at the specific insights from our case confirming the criticality of the respective success factors. Considering methodological, organizational, socio-cultural, and technical aspects, we have grouped the critical success factors into five management dimensions as shown in Table 1, and for each focusing on the Design Thinking key principles, user-centricity and co-creation. These are: (1) Stakeholder Relationship Management, (2) Objectives Management, (3) People Management, (4) Project Management, and (5) Solution Consistency Management. They are all affected by and in return also influence leadership. The table includes guiding questions for each critical factor and the associated major lessons learned of each phase.

- *Stakeholder Relationship Management* is the most critical success factor and needs continuous highest consideration along the complete project lifecycle. In general, it is necessary to understand the larger project frame and to find allies to reach commitment, namely for funding and resources. With respect to Design Thinking, they need to endorse and adhere to the key principles of user-centricity and co-creation. This includes top-level sponsors, problem owners, and user representatives. It requires a strong and ambidextrous leadership (Odgers Berndtson 2017) to unite all interests and keep them aligned to the common purpose of the project and specifically keep focus on user needs and desires. Quick wins and early tangible results help to enforce the willingness of operations to participate in the generation of user insights and help to create the necessary evidence of positive impact to keep also investors on board. The larger your alliance for the project the better it can overcome blocking points and resist political agendas/concerns and stakeholder conflicts, when they surface at a later stage. First, it is necessary to onboard at least one initial investor, who pays and provides resources for the project. In large organizations the accountability for planning investments (budget) and the responsibility to execute them operationally (namely resources assignment) are different functions. The second important stakeholder group are the developers, who make the prototypes and finally develop the new solution. The third, probably most difficult and with respect to Design Thinking most critical stakeholder group consists of the users, who express their needs and who will ultimately have to use the new solution and then benefit from it. We also included the works council as the formal user representatives, who is a natural ally for employee engagement and therefore Design Thinking. They helped us a lot to open doors and to increase project reach.
- *Objectives Management*, the second dimension encompasses the success factors strategy-fit, framing, and focus. Why is the project important to the company and to users? What is the proposed value or impact created by the project and which concrete problem does it solve? The Design Thinking project should contribute solving complex problems linked to a top-level company objective. Doing so, it visibly contributes to the success of the company and can more easily increase its own awareness. But this only works if the project shows evidence on how it helps to solve a concrete operational problem, to get the job done for end users, in our

Table 1 Critical success factors and associated insights

Critical Success FACTOR	Guiding QUESTIONS	Phase 1 INSIGHTS	Phase 2 INSIGHTS	Phase 3 INSIGHTS
(1) STAKEHOLDER RELATIONSHIP MANAGEMENT	Who are the key stakeholders?	Build an alliance for the project including top level sponsors, the problem owner, and user representatives.	Make sure that overall funding and resources are committed upfront; if not continuation of the project is at risk.	Try to keep a stable core team andavoid a complete handover at later stage.
Commitment for Funding and Resources	Who provides funding? Who is developing the solution? Who needs and will use it?	Understand and build trust in the larger project frame, i.p. with the actual development and the responsible operations team	Deliver quick wins and tangible results to trigger the interest of operations and to keep investors on board.	... if not the risk of delay, termination,or diversion of the project is very high.
(2) OBJECTIVES MANAGEMENT	Why is the project important to the company and to the users?	Set the focus on the company top level objectives and show how you contribute to solving a concrete problem.	Be ready to pivot and to ensure that sponsors and investors are wiling to deal with uncertainty and the ambiguity	Prepare for scaling when choosing a launching customer.
Strategy Fit, Framing & Focus	What is the value or impact created by the project? Which problem does it solve?	Integrate the Design Thinking initiative into a lighthouse project with top level priority.	Outcomes of exploration during the prototyping and testing phase are unknown	Deal with political concerns upfront; they will be harder to overcome when they surface at a later stage.
(3) PEOPLE MANAGEMENT	Which specific expertise, skills and competency do you need?	Rely on a cross-functional team built of passionate T-shaped characters.	Increase domain knowledge; maintain the ability to involve additional users and domain experts.	Anticipate co-location and protect project core team and lead users when changing to a new set-up.
Teaming, Skills & Competency	Which expertise is available in-house, which requires external support?	Focus initially on a very high level of method knowledge to perform need finding and understand users	Focus on method knowledge for fast prototyping and testing; if required onboard specialized partners.	Focus on the involvement of solution owners and operations.
(4) PROJECT MANAGEMENT	Can Design Thinking be integrated in the project structure? How can different approaches be managed?	Define a dedicated subproject or work stream and integrate it in a lighthouse project; synchronize milestone plans; keep full accountability for deliverables.	Be agile: Go fast, Co-create, Iterate, and dare to pivot.	Keep the end-to-end perspective in mind and adapt it to any upcoming change.
Planning, Alignment & Synchronization	Is there dedicated space available?	Reserve a dedicated space for co-creation.	Adjust skills and resource needs dynamically depending on changing levels of expected outcome.	Manage expectations to ensures a smooth transition from phase to phase.
(5) SOLUTION CONSISTENCY MANAGEMENT	Is the solution compliant with the existing architecture and can it be seamlessly integrated?	Emphasize user journeys and user stories as means to manage detailed requirements and to reach acceptance for new digital solutions.	Agile prototyping leads to standalone and often disjunctive prototypes. Ensure interfaces (API) are available to connect prototypes where-ever possible to the operational systems.	Get clearance about change potential at the start of the project. At imple-mentation it is too late and too costly.
IT Architecture & Infra-structure, Data Mgmt., Privacy & Data Security	How can prototypes be made functional with real data?	Get to know the bigger picture and be aware of legacy concerns. Respect the current standards or be authorized for defining the new standards.	Anticipate privacy and data security issues; they need to be considered when using real production data.	IT will discard any new "standard" or principal design or architecture change at implementaton stage.

case reducing lead time and cost of non-quality. In addition, the Design Thinking project must be ready to pivot while keeping the objectives aligned and must ensure that sponsors and investors are willing to deal with uncertainty and literally "dance with ambiguity" (Leifer and Steinert 2011) of the unknown outcomes of exploration.

- *People Management* is probably the most important dimension for keeping the project running. Leadership and the constitution of the team make the difference. Who are the right people able to form a Design Thinking project team? Which specific expertise, skills, and competency does the project need? Which expertise is available in-house, which requires external support? Look for a great leader and passionate people. For Design Thinking, rely on a cross-functional team built of T-shaped personal profiles and utmost important, maintain the ability to continuously involve users, experienced operational people as well as newcomers, and recognized domain experts. At the beginning of the project, end users and their operational management were very skeptical but with each prototyping cycle we could increase user engagement and participation. Proficiency in Design Thinking methods is the baseline and starting point. The need for operational domain knowledge increases over time and is subsequently enriched by domain specific industrialization and integration capabilities. The needs change with each phase and require a dynamic adaptation of skills and resources depending on changing levels of expected outcome. In consequence, the project must be ready to add new team members and to anticipate eventual changes of key partners. For example, in our use case, on the one hand, we successfully added digital design skills enabling digital prototyping and agile co-creation in Phase 2. On the other hand, the change of the team in Phase 3, new prime contractor and development partner, led to the discontinuation of the project. The missing involvement beforehand provoked a "not-invented-here" attitude of the new development team and with the change of site, we lost a large majority of the users. We learned that it is important to protect the project core team and to avoid a hand over and change of responsibilities midway.
- *Project Management* organizes the way of working together, risk assessment and decision making, the planning of resources, milestones, and deliverables. Again, here the project leader is the key person to balance interests orchestrating the team. The question is how to integrate a Design Thinking project in a larger project structure when cultures clash and as in our case we have to deal with potentially opposing ways of working between an agile and a classical waterfall approach? By defining a dedicated project structure for Design Thinking and integrating it as an independent work stream in one of the company's lighthouse projects, we were able to align the formal milestone plans while keeping accountability for our deliverables, maintaining the necessary speed and flexibility for our approach. This helped to manage expectations and to ensure a smooth transition from phase to phase. We missed to anticipate the radical handover in Phase 3. But late changes if not anticipated represent a high risk of failure resulting in delay, termination, or diversion of the project. In our case, new sponsors changed priorities. The new development team rejected "foreign" change requests since

the detailed specification of the targeted solution was already finalized and therewith our opportunity window was closed. An additional success factor is the availability of an open dedicated project space big enough for the collocation of the complete project team, with enough room for extensive prototyping and testing, and with the flexibility to change the space setup depending on changing project needs. At best, the space should be in proximity to the users. The users will not follow the project, the project has to come to the users. Thus, with the change of the project location in Phase 3, it was very difficult to maintain the contact with the users that had previously worked with the project on the local site.
- *Solution Consistency Management* bundles a larger number of associated success factors and in particular, considering software-intensive products and services, many technical aspects such as data management, infrastructure and network, system architecture and integration, as well as data security and privacy policies. Is the underlying IT architecture fit for a seamless integration? Design Thinking requires prototypes that can be made functional with real data. Here, standalone disjunct prototypes are the easiest way to gather user feedback in a short time frame. The drawback of this approach is that privacy and data security issues surface when connecting to real production data. We had application programming interfaces (API) proved to be workable but albeit the prototypes were fully functional and based on state-of-the-art design choices, we did not connect them to systems in operation. To ensure compliance and consistency of the future solution, it is important to understand the bigger picture of the global IT system architecture and infrastructure, protocols, and standards. The Design Thinking project has to get clearance about the change potential and possible legacy concerns at the start of the project, because very often they represent the biggest and remaining obstacle for change. At implementation it is too late and too costly for principal design or architecture changes.

Finally, let us consider the lessons learned that might be applied to other similar projects. In the following section, we present the results of our SWOT analysis for the introduction of Design Thinking in a large manufacturing organization.

- *Strengths* of the approach are user acceptance and high adoption rates due to the involvement and participation of users during the entire process. The approach helps to create a holistic view from a user perspective. Co-creation unlocks tacit user knowledge and creates in-depth problem understanding. Fast iterations of prototyping and testing with a focus on minimum viable functions make outcomes tangible in various maturity levels enabling direct user feedback and validation until a good and effective, usable and useful, new solution is created.
- *Weaknesses* however the explorative nature of Design Thinking makes it difficult to plan required resources upfront, as these are dependent on the outcomes of the process. In particular, the high efforts required for need finding and problem understanding are underestimated, as solutions for obviously apparent problems are demanded leading to difficulties for recruiting end users, who are normally assigned to operational priorities. Investors are traditionally solution-driven

focusing on optimization and efficiency and are not well suited to deal with uncertainty of a user-centered problem-driven Design Thinking approach focusing on impact and effectiveness.

- *Opportunities* are employee engagement and trust building by involving all stakeholders and in particular end users in co-creation and collaborative design, namely in large cultural or digital transformation initiatives. Valuable user insights increase the effectiveness of the design. The application of Design Thinking results in a strong bottom-up sense of ownership for the solution (s) developed and increases user acceptance. Integrated in a lighthouse project, Design Thinking can more easily adopted by the organization becoming an essential part of the company's "DNA," the global product and service development, most critical for software-intensive solutions.
- *Threats* are the fast pace of the Design Thinking project, thus the synchronization with other initiatives may not be sufficient, particularly if the project teams are not available for and not used to agile co-creation. Also, the integration with legacy systems may demand more time and incremental focus. Lastly, by the threat of not considering alignment and synchronization with parallel or superordinate projects, a strong not-invented-here reflex can be triggered when eventually merged at a later stage.

Conclusion: Key Takeaways and Lessons Learned

With our experience gained in this concrete case, we reached a better understanding of how Design Thinking can be applied for a better design of complex industrial processes and manufacturing organizations. By involving users and utilizing their tacit knowledge we could increase awareness about key challenges and set focus on the real problems, things that really matter. By iteratively making progress tangible through prototypes, knowledge was made explicit and gathered beyond the incremental. However, in order to use Design Thinking in a large corporation, it is necessary to stress stakeholder management. Finally, the speed and overarching approach of the work stream led to issues that several parties felt not to be involved adequately and definitely too late. It became obvious that it is not enough to include the otherwise most of the time forgotten end users—the ones who need—but also to include the solution developers—the ones who make—and last but not least the business owners and investors—the ones who pay—in a holistic user-centered approach.

We were able to demonstrate how Design Thinking accelerates and improves the usability and effectiveness of design, in our case of a smart, collaborative support system for workers on the shop floor. We learned that there is no general recipe. Ambidextrous leadership, perseverance and resilience in keeping end-to-end accountability, and a collaborative mindset are necessary to make Design Thinking a success. The simplification of our approach in three subsequent phases helped a lot to make it sizable for a large organization. Overall, Design Thinking is a very suitable approach for developing a profound problem understanding as a source for innovation, namely in the design of software-intensive products and services.

References

Abrell T, Durstewitz M (2016) The role of customer and user knowledge in internal corporate venturing: the viewpoint of the corporate entrepreneur. Int J Technol Manag 71(3/4):171–185
Boy GA (2011) Cognitive function analysis in the design of human and machine multi-agent systems. In: Boy GA (ed) The handbook of human-machine interaction: a human-centered design approach. Ashgate Publishing, Surrey, pp 189–206
Brenner W, Uebernickel F, Abrell T (2016) Design thinking as mindset, process, and toolbox. In: Brenner W, Uebernickel F (eds) Design thinking for innovation. Springer, Heidelberg, pp 3–21
Brown T (2008) Design thinking. Harv Bus Rev 86(6):84–92
Brown T (2009) Change by design. HarperCollins, New York
Design Council (2019) The Double Diamond: 15 years on. Design Council. https://www.designcouncil.org.uk/. Accessed 11 Jan 2021
Durstewitz M (1994) Reuse of experience for the design of industrial applications - a formal approach. In: Barthélemy J, Bisdorff R (eds) Cognitive science in industry: proceedings of the first European Conference on Cognitive Science in Industry, Luxembourg, 28–30 September
Grudin J (1993) Obstacles to participatory design in large product development organizations. In: Schuler D, Namioka A (eds) Participatory design: principles and practices. Lawrence Erlbaum Associates, Hillsdale, NJ, pp 99–119
Hobday M (2000) The project-based organisation: an ideal form for managing complex products and systems? Res Policy 29(7-8):871–893
Hollnagel E (2009) The ETTO principle: efficiency-thoroughness trade-off: why things that go right sometimes go wrong. CRC Press, Boca Raton
IDEO (2015) Design kit: bundle ideas. IDEO.org. https://www.designkit.org/methods/30. Accessed 18 Jan 2021
Issar G, Navon LR (2016) Operational excellence - a concise guide to basic concepts and their application. Springer, Cham
LEGO Group (2010) Lego serious play open source. Introduction to LEGO serious play. https://seriousplaypro.com/about/open-source/. Accessed 11 Jan 2021
Leifer L, Steinert M (2011) Dancing with ambiguity: causality behavior, design thinking, and triple-loop-learning. Inf Knowl Syst Manag 10:151–173
Odgers Berndtson (2017) Ambidextrous leadership. Manager Barometer 2017/2018, pp 38–42. https://www.odgersberndtson.com/media/5652/ob_manager_barometer_2017.pdf. Accessed 12 Feb 2021
Ries E (2011) The lean startup. Penguin Books, London
Stickdorn M, Schneider J (2010) This is service design thinking, basics - tools – cases. BIS Publishers, Amsterdam
Ulwick AW (2016) Job to be done: theory to practice. Idea Bite Press, Chicago
von Hippel E (1994) “Sticky information” and the locus of problem solving: implications for innovation. Manag Sci 40(4):429–439

Digital Platform Design at the Edge of Complexity: The Value of Design Thinking to Balance Between Configuration and Customization

Emanuel Stoeckli

Introduction

Design-orientation has become an essential factor to create unique value propositions that enable higher margins, faster growth, and better organizational performance (Rae 2015). In fact, organizations with a top-quartile McKinsey Design Index grow twice as much compared to the industry-benchmark (Sheppard et al. 2018). As such, human-centric design approaches such as Design Thinking gained substantial traction in the development of software-intensive products and services. The nature of today's products and services, however, is shifting toward digital platforms. The most valuable companies are platform-based and the number of digital platforms is continuously growing (Cusumano et al. 2019; Evans and Gawer 2016). A distinct characteristic of platforms is that they bring together multiple sides of actors. Hence, the needs, concerns, values, and perceptions of various involved users and stakeholders need to be balanced—a task which is known to be key pillar of human-centered design approaches (Hehn et al. 2020; Rouse 2007). For example, Design Thinking artifacts such as prototypes can be used to explicate different perceptions and conflicting interests, hence, may help to manage tensions that emerge between stakeholders (Staehelin et al. 2021). However, despite these benefits of human-centric approaches their focus lies on user requirements while technical system requirements are neglected (Hehn et al. 2018b) and there is a lack of research on their application to the design of digital platforms.

This gap is crucial, since the design of digital platforms differs from other products and services in that it is thoroughly based on modularity to increase variety and reduce complexity (Gawer 2014). Hence, the very core of platform design is the systematic reuse of modules. This needs to be considered when triangulating between desirability, viability, and feasibility. Conversely, satisfying the

E. Stoeckli (✉)
esurance AG, Zürich, Switzerland

J. Hehn et al. (eds.), *Design Thinking for Software Engineering*, Progress in IS,
https://doi.org/10.1007/978-3-030-90594-1_12

(conflicting) needs of various stakeholders requires a high degree of individual customization. In turn, it leads to a high degree of complexity and a low degree of reusability, which is against the core principle of digital platforms. In contrast, pure maximization of reusability by configuring existing platform modules in the hope to maximize economies of scope and scale is far away from human-centered design, neglects user needs, and exhibits solution-fixation (Meinel and Leifer 2019). Therefore, this contribution triangulates the two perspectives and discusses the value of Design Thinking as a mindset, process and toolset for Digital Platform Design.

Theoretical Foundation

Platform Types: On Seeing the Forest for the Trees

Literature emphasizes the distinction between *transaction platforms* and *innovation platforms* (Cusumano et al. 2019; Evans and Gawer 2016). While the former facilitates transactions between two or more actors, the latter facilitates the creation of product and service innovations based on shared resources (Cusumano et al. 2019). As such, platforms per se exist for a long time. For example, physical platforms bring together people to access shared resources like trains, while product platforms harness shared resources to create novel products within a common product family (e.g., car models). In this contribution, we focus on *digital platforms* that benefit from both types. In reality, they often act as *integrated platforms* that combine an innovation platform (e.g., app development) together with a transaction platform (e.g., app marketplace) (Evans and Gawer 2016). Digital transaction platforms benefit from lower transaction costs and exponential scalability (Hein et al. 2020). Digital innovation platforms benefit from digital (or digitally equipped) products and services that embrace characteristics such as being (re-)programmable, addressable, sensible, communicable, memorable, traceable, and associable (Yoo et al. 2010). The crux is that these characteristics enable flexibility and openness, which in turn yields in generativity and digital convergence (Yoo et al. 2012). Generativity implies that the form and function of the facilitated innovations becomes inherently dynamic and malleable so that a wide and unforeseen variety of forms and functions can be (re-)configured without need for prior customization. Convergence implies that the scope of digital platforms is not limited to individual industries and domains anymore, but functionalities from previously separate disciplines are converging.

Platform Economics: On Economies of Scale and Scope

A central success factor of digital platforms is their tremendous *network effects* which lead to winner-take-all markets (Constantinides et al. 2018; Katz and Shapiro 1985). Specifically, the value of a platform scales with (and depends on) the number of actors within one side of the platform (direct network effect) as well as across multiple sides of the platform (indirect network effects) (Evans and Gawer 2016). This is why platforms widely face "chicken and egg" problems to attract actors (Caillaud and Jullien 2003). Apart from *economies of scale*, the major benefit of innovation platforms lies in realizing *economies of scope* through the systematic reuse of shared modules (i.e., components) for the creation of new complementary products and service innovations (i.e., complements) (Gawer 2014). In addition, the value of innovation platforms, then, scales with the number (and quality of) of reusable components and complements (Cusumano et al. 2019). Accordingly, opening up innovation platforms builds the foundation to create digital platform ecosystems that attract complementors such as developers (Hein et al. 2020). Consequently, the value creation logic in digital platform ecosystems shifts toward mutual value co-creation between the actors (Hein et al. 2019). *Governance structures*, then, control the access of actors to resources and regulate the interactions and value allocation between these actors through price and incentive mechanisms (Boudreau and Hagiu 2009).

Platform Modularity: On Complexity and Variety

The described economies of scope and scale draw on the principle of *modularization*. Digital platforms are modular if their structure (i.e., architecture) is decomposed into smaller building blocks (i.e., modules, components) with well-defined functions and relations (i.e., interfaces) so that the components can be operated on independently and the initially complex system becomes manageable (Baldwin and Clark 2000; Bask et al. 2010). From that, two key benefits arise from modularity:

First, the *complexity of a platform is reduced* by limiting the scope of interaction to a set of well-defined interfaces between a limited number of components (Baldwin and Clark 2000). This abstraction allows to isolate components so that complex parts can be hidden from other components (Viana et al. 2017).

Second, the *variety of a platform is enhanced* by increasing the flexibility of components through basic operations such as splitting, substituting, augmenting (including), excluding, inversion, and porting (Baldwin and Clark 2000). Hence, a tremendous potential for unique product and service innovations is created at each level of decomposition (i.e., modules or components) (Germonprez et al. 2007). Thereby, platform owners often introduce different types of modules to control the possible basic operations. For example, core modules which can be extended with optional modules (Hein et al. 2020), and unique modules in addition to common and variant modules (Moon et al. 2010).

Platform Design (Thinking): On Configuration and Customization

Modularity enables to adapt to diverse and continuously changing user needs while realizing efficiencies of scale and scope (Bask et al. 2010; Brax et al. 2017; Moon et al. 2010; Yoo et al. 2010). To do so, two ways are distinguished: *Configuration* supports variance through setting predefined parameters, while customization supports meeting individual requirements that go beyond the configurable limit (Sun et al. 2008). Accordingly, a trade-off becomes apparent: either the satisfaction of individual needs and requirements is restricted to predefined configurations or complexity is increased by introducing customized or customer-specific modules (Brax et al. 2017). It is here where human-centered design methods come into place to find the right balance between desirability, viability, and feasibility (Cooper et al. 2007).

Approach

This contribution draws on digital platform and design thinking literature to triangulate theory with the design activities made in practice while acting as Platform Product Manager of a B2B2C InsurTech platform at esurance in Switzerland. The reflection process grounds in experiences made in the time frame between August 2019 and January 2021. The platform itself has to be seen in the context of the insurance industry in which value is increasingly created in value networks (Stoeckli et al. 2018). The platform brings together insurance companies, associations, and other stakeholders relevant for different target groups (e.g., Gastronomy, IT, insurance company-specific white label solutions) to co-create digital insurance platforms. As an *innovation platform*, the B2B2C platform facilitates innovation in form of the creation of new complementary transaction platforms with a variety of configuration possibilities related to products (e.g., tariffs), pre-sales services (e.g., leads), sales services (e.g., partner app), and post-sales services (e.g., invoice generation). From a *transaction platform* view, a variety of target group specific platforms facilitate the digital exchange of insurance products and services in the context of small and medium-sized enterprises (SMEs). For example, transactions such as calculating digital offers, buying insurance online with instant temporary coverage, or accessing previous transactions in a Cockpit. This case is particularly interesting to investigate the phenomena at hand, since in comparison to product-oriented insurances, the platform adopts an inherently customer-centric approach (i.e., variety of platforms with products and services most relevant to stakeholders from a particular target group).

The Value of Design Thinking for Digital Platform Design

Mindset: Increasing Variety of Addressed Problems per Solution

Successful product and service innovations—as proposed by human-centered approaches like design thinking—combine desirability, feasibility, and viability (Cooper et al. 2007; Jones and Samalionis 2008).

Design Thinking starts with desirability and with a problem-oriented mindset (Meinel and Leifer 2019). Digital Platforms, in turn, can be seen as a class of solutions that harness modularity (feasibility) to gain economies of scope and scale (viability). Consequently, a digital platform mindset is at risk of being biased toward re-using existing platform capabilities rather than exploring the problems of users. This risk is widely known as solution-fixation (Meinel and Leifer 2019). Design Thinking fosters a deep and broad understanding of the problems and prevents bias toward viability and feasibility (Jones and Samalionis 2008). Hence, the variety of considered problems increases. On the other hand, Design Thinking often results in finding particular solutions to particular problems (Dolata and Schwabe 2016) and ends with building nontechnical prototypes. Hence, the risk is that the innovation process is biased toward individual desirable solutions. From a platform point of view, individual solutions may be feasible in the short term, but the logic of platform viability lies in satisfying individual user needs through configurations.

Therefore, I argue that the sweet spot is where a wide range of relevant problems are addressed by a variety of platform configurations (i.e., few solutions) (see Fig. 1).

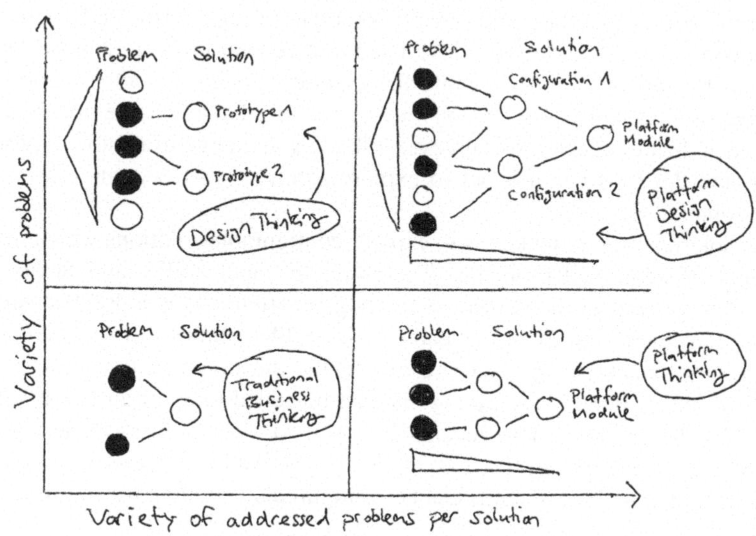

Fig. 1 Increasing variety of problems and variety of addressed problems per solution

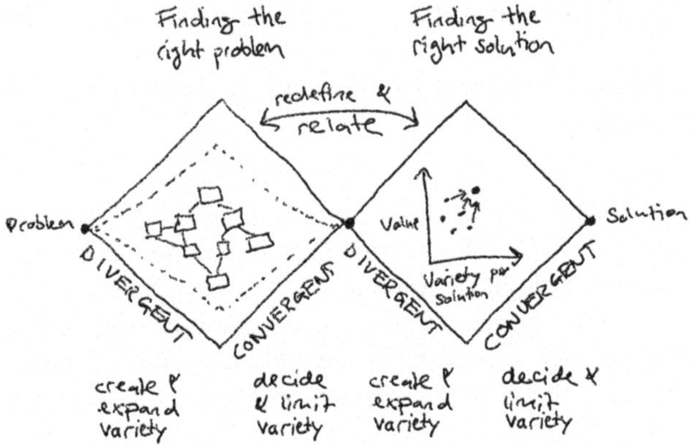

Fig. 2 Double Diamond of divergent and convergent phases (based on Design Council 2015)

Therefore, the divergent and convergent thinking of Design Thinking must be combined with the platform striving to address many problems with a few configurable solutions.

In the problem space, Design Thinking helps to continuously redefine meaningful problems for relevant stakeholders. Systematic methods such as "design by analogy" WordTrees help to identify relevant analogies and problems by mapping more general and specific terms at various abstraction levels (Linsey et al. 2012). However, today's problem space is shaped by (limitations of) the existing solutions. Hence, it is crucial to incorporate how the identified problems relate to problems already addressed by the platform without jumping to the solution space. For example, imagine the context of legal terms that need to be accepted by users. One way to look at it is to explore the specific problem of legal terms and the trade-off between completeness and readability. The more general problem is how complex issues can be shown concisely while the complex details remain available only where they are needed. As such, links to related problems become apparent (e.g., definitions in health declaration surveys). Like this the range of identified problems expands and the foundation is set to explore solutions that become relevant in a broader context.

In the solution space, the key is to explore configurable solutions which increase the variety of addressed problems. Therefore, the malleability and flexibility of digital solutions can be harnessed to foster generativity. It is helpful to build on ideas and turn them into more malleable and flexible solutions that tackle multiple identified problems (see Fig. 2). For example, the illustrated issue with legal terms was solved with a flexible tooltip tag that can be added to any translation which is rendered by the frontend. This means it is open to a yet unforeseeable variety of contexts (e.g., legal terms in a basket, price calculation parameter descriptions, survey questions) and translations can be configured at any time by non-developers.

Process: Considering the Lifecycle and Setting Boundaries

From a process perspective, Design Thinking adopts a micro cycle that includes re-defining the problem, need finding, ideation, prototyping, and testing (Brenner et al. 2016). While the phases may vary, the general logic is to alternate between divergent and convergent phases as well as between the problem and the solution space as depicted in Fig. 2 (Design Council 2015). This micro cycle is repeatedly applied so that the understanding of both spaces continuously increases. Structuring multiple iterations of micro cycles into an overall innovation process is referred to as macro cycle. As such, the macro cycle can again be divided into a divergent and convergent phase so that the fidelity of prototypes gradually increases with every iteration from low-fidelity prototypes such as paper prototypes to high fidelity prototypes such as clickable mockups (Brenner et al. 2016). Applied to digital platform design, this contribution highlights two learnings:

First, in course of transitioning iteratively from prototypes toward actual implementations, three paths of deriving human-centric requirements can be taken:

1. *Platform Configuration*: Deriving human-centric requirements on how to reuse existing platform modules, components, and features.
2. *Platform Customization*: Deriving human-centric requirements on how to create new custom modules, components, and features.
3. *Platform Expansion*: Deriving human-centric requirements on how to extend the platform by a) altering and enriching the configuration capabilities of existing modules, or b) by adding new configurable modules, components, or features.

However, in practice, innovations often comprise a mixture of these three paths, which demands for a careful balancing between configuration and customization. Thereby two main challenges need to be addressed. On the one hand, there is always a trade-off between short-term needs of individual customers and long-term needs such as maintainability, scalability, and extensibility of the platform (De Weck et al. 2011). On the other hand, the challenge lies in distinguishing individual customization needs that occur today from needs of multiple future users that demand for configurability. Design Thinking artifacts such as personas can help to approach this challenge, for example, by defining primary and secondary personas as well as anti-personas. This is especially true in earlier stages of digital platform design where the target users and the direction in which the platform should be expanded are less steady. Nevertheless, a clear vision is needed on where the platform aims to realize economies scope in the future. In fact, it is the very essence of strategy is to decide what to do and what not to do to build unique value propositions (Porter 1996, 2012).

Second, it is crucial to understand that innovation happens at each level of modular decomposition (e.g., modules, components, features) and different modules are in different lifecycle phases (e.g., a well-established core module versus a novel complementary module). Like teams that transition through forming, storming, norming, and performing stages (Tuckman 1965) modules evolve over time. This is true not only true from a technical component view, but for the whole socio-technical system. Specifically, not only modules may be new, but also new users with different needs may start to engage with an existing module. Hence, both the problem and the solution space continuously evolve over time. Long-term success—as proposed by ambidexterity theory—grounds in the ability to simultaneously explore (i.e., search and experiment by increasing variety) and exploit (i.e., increase productivity and efficiency by limiting variety) (March 1991). In analogy, car manufacturers only create full redesigns of car models every 4–6 years, while innovation in between is limited to facelifts. Accordingly, incremental, evolutionary, and revolutionary innovations should be pursued simultaneously, which is possible with Design Thinking (Brown 2009). For the design of digital platforms, the challenge is to manage a portfolio of innovations as part of a hierarchical decomposition of modules, components, and features. When aiming at revolutionary innovations, it is important to start as open minded as possible into the design process and nothing changes from an ordinary design process. On the contrary, when aiming at more incremental and evolutionary innovations it is necessary to link the design process to the related modules. On the one hand, each module has an underlying value creation logic and solves actual problems of users. On the other hand, the boundaries need to be defined by deciding carefully to what extent the design space should be limited (and extended) while also taking into consideration the risk of solution-fixation.

Toolset: Combining Design Methods with Modular Operators

The vast body of methods and tools used in design (thinking) are solution agnostic and can, thus, be applied to digital platform design as well (Hanington and Martin 2012; IDEO 2015; Uebernickel et al. 2015). The particularities of platform design thinking do not lie in the methods and tools, but in the way they can be applied to platforms. Prior research has elaborated on the usefulness of different methods along the design process (Hehn et al. 2018a). When applying these design methods to digital platform design, the main particularities lie in the prototyping and testing phase. Namely, the modular basis operations of splitting, substituting, augmenting (including), excluding, inversion, and porting (Baldwin and Clark 2000) simplify the transition from low-fidelity to high fidelity prototypes. For example, substitution allows us to test a new microservice that generates PDF documents on a single platform instance.

Concluding Remarks

This contribution has analyzed the value of Design Thinking for the design of digital platforms. From a mindset view, the contribution proposes to combine the divergent and convergent thinking of Design Thinking with the platform mindset of addressing many problems with a few configurable solutions. In the context of digital platforms, the sweet spot of desirable, feasible, and viable solutions is where a wide range of problems are addressed by a variety of platform configurations. From a process view, this contribution underlines the strategic relevance to carefully balance between configurations and customizations and highlights the challenge of distinguishing between today's customization needs and future configuration needs. From a toolset view, the contribution suggests that existing design methods and tools together with the modular basis operations simplify the transition toward high fidelity prototypes. With this discussion, the contribution provides a better understanding of how Design Thinking can be integrated into the ongoing design and development of digital platforms.

References

Baldwin CY, Clark KB (2000) Design rules: the power of modularity. MIT Press

Bask A, Lipponen M, Rajahonka M, Tinnilä M (2010) The concept of modularity: diffusion from manufacturing to service production. Int J Manuf Technol Manag 21(3):355–375

Boudreau KJ, Hagiu A (2009) Platform rules: multi-sided platforms as regulators. Platf Mark Innov 1:163–191

Brax SA, Bask A, Hsuan J, Voss C (2017) Service modularity and architecture – an overview and research agenda. Int J Oper Prod Manag 37(6):686–702

Brenner W, Uebernickel F, Abrell T (2016) Design Thinking as mindset, process, and toolbox. In: Brenner W, Uebernickel F (eds) Design Thinking for innovation: research and practice. Springer, Cham, pp 3–21

Brown T (2009) Change by design: how design thinking transforms organizations and inspires innovation. Harper Business, New York

Caillaud B, Jullien B (2003) Chicken & egg: competition among intermediation service providers. Rand J Econ 34(2):309–328

Constantinides P, Henfridsson O, Parker GG (2018) Introduction—platforms and infrastructures in the digital age. Inf Syst Res 29(2):381–400

Cooper A, Reimann R, Cronin D (2007) About face 3: the essentials of interaction design. John Wiley & Sons

Cusumano MA, Gawer A, Yoffie DB (2019) The business of platforms: strategy in the age of digital competition, innovation, and power. Harper Business, New York

De Weck OL, Roos D, Magee CL (2011) Engineering systems: meeting human needs in a complex technological world. MIT Press

Design Council (2015) What is the framework for innovation? Design council's evolved double diamond. Design Council. https://www.designcouncil.org.uk/news-opinion/what-framework-innovation-design-councils-evolved-double-diamond. Accessed 7 Feb 2021

Dolata M, Schwabe G (2016) Design thinking in IS research projects. In: Brenner W, Uebernickel F (eds) Design thinking for innovation. Springer, Cham, pp 67–83

Evans PC, Gawer A (2016) The rise of the platform enterprise: a global survey. The Center for Global Enterprise. http://epubs.surrey.ac.uk/811201/

Gawer A (2014) Bridging differing perspectives on technological platforms: toward an integrative framework. Res Policy 43(7):1239–1249

Germonprez M, Hovorka D, Collopy F (2007) A theory of tailorable technology design. J Assoc Inf Syst 8(6):315–367

Hanington B, Martin B (2012) Universal methods of design: 100 ways to research complex problems, develop innovative ideas, and design effective solutions. Rockport Publishers, Massachusetts

Hehn J, Uebernickel F, Herterich M (2018a) Design thinking methods for service innovation—a Delphi study. In: Proceedings of the 22nd Pacific Asia Conference on Information Systems, Yokohama

Hehn J, Uebernickel F, Stoeckli E, Brenner W (2018b) Designing human-centric information systems: towards an understanding of challenges in specifying requirements within design thinking projects. MKWI proceedings 2018

Hehn J, Mendez D, Uebernickel F, Brenner W, Broy M (2020) On integrating design thinking for human-centered requirements engineering. IEEE Softw 37(2):25–31

Hein A, Weking J, Schreieck M, Wiesche M, Böhm M, Krcmar H (2019) Value co-creation practices in business-to-business platform ecosystems. Electron Mark 29(3):503–518

Hein A, Schreieck M, Riasanow T, Setzke DS, Wiesche M, Böhm M, Krcmar H (2020) Digital platform ecosystems. Electron Mark 30(1):87–98

IDEO (2015). Field guide to human-centered design. IDEO.org. https://www.designkit.org/resources/1. Accessed 6 Feb 2021

Jones M, Samalionis F (2008) From small ideas to radical service innovation. Des Manag Rev 19(1):20

Katz ML, Shapiro C (1985) Network externalities, competition, and compatibility. Am Econ Rev 75(3):424–440

Linsey JS, Markman AB, Wood KL (2012) Design by analogy: a study of the wordtree method for problem re-representation. J Mech Des 134(4):041009

March JG (1991) Exploration and exploitation in organizational learning. Organ Sci 2(1):71–87

Meinel C, Leifer L (eds) (2019) Design thinking research: looking further: design thinking beyond solution-fixation. Understanding innovation. Springer, Cham

Moon SK, Shu J, Simpson TW, Kumara SRT (2010) A module-based service model for mass customization: service family design. IIE Trans 43(3):153–163

Porter ME (1996) What is strategy. Harv Bus Rev 74(6):61–78

Porter ME (2012) What is strategy: issues for the world bank. Harvard Business School. https://www.hbs.edu/faculty/Publication%20Files/2012-0802—World_Bank_Strategy_c2726162-7d36-400b-938c-a87119f5ccac.pdf. Accessed 26 Jan 2021

Rae J (2015) Design value index. Des Manag Rev 26(1):4–8

Rouse WB (2007) People and organizations: explorations of human-centered design. Wiley

Sheppard B, Sarrazin H, Kouyoumjian G, Dore F (2018) The business value of design. McKinsey Quarterly

Staehelin D, Dolata M, Schwabe G (2021) Managing tensions in research consortia with design thinking artifacts. In: Hehn J, Mendez D, Brenner W, Broy M (eds) Design thinking for software engineering - creating human-oriented software intensive products and services. Springer

Stoeckli E, Dremel C, Uebernickel F (2018) Exploring characteristics and transformational capabilities of InsurTech innovations to understand insurance value creation in a digital world. Electron Mark 28(3):287–305

Sun W, Zhang X, Guo CJ, Sun P, Su H (2008) Software as a service: configuration and customization perspectives. In: 2008 IEEE Congress on Services Part II (services-2 2008)

Tuckman BW (1965) Developmental sequence in small groups. Psychol Bull 63(6):384–399

Uebernickel F, Brenner W, Naef T, Pukall B, Schindlholzer B (2015) Design thinking: Das Handbuch. Frankfurter Allgemeine Buch

Viana DD, Tommelein ID, Formoso CT (2017) Using modularity to reduce complexity of industrialized building systems for mass customization. Energies 10(10):1622

Yoo Y, Henfridsson O, Lyytinen K (2010) Research commentary—the new organizing logic of digital innovation: an agenda for information systems research. Inf Syst Res 21(4):724–735

Yoo Y, Boland RJ, Lyytinen K, Majchrzak A (2012) Organizing for innovation in the digitized world. Organ Sci 23(5):1398–1408

Design Thinking in Healthcare—Enabler for Digitalization in Complex Environments: Why Healthcare Is Adequate to Proof the Potential of Design Thinking for Software-Intensive Ecosystems

Christophe Vetterli

Two Different Worlds

Let us step back one step first: Currently, most patient experiences look like the following emergency example: You are feeling some unusual pain in the belly region for 2 days, and you have googled several sources. You are waiting for one more day, with uncertainty because you do not know if you are right from your googled deviated diagnosis, to then decide to look for the emergency department (ED) from the regional hospital. Before you could enter ED, you have parked in the far distanced parking because the six parking lots directly in front of the entrance were taken. You now head to the receptionist, who tells you to fill out the standardized form. The first five items are name, birthday, civil status, religion, and insurance status. The latter three do not really match your current needs, and unfortunately, you have been here 3 weeks ago and already provided that information—nothing has changed since then. However, you need to fulfill the data again. Your need seems "a classic" for an ED entry: You want to know what is going on with your belly. After providing data—again—you start waiting in the waiting area, having the first contact to an assistant physician after 50 min arriving at the ED. He has left you in the examination room after 8 min telling you that the senior physician will need to have a look. To shorten the story, you will wait on average another 30–60 min, depending on weekday and daytime, to have full transparency of the next treatment options. Typically, this is still not the end of your ED stay. Central Emergency Departments in hospitals have between 20,000–80,000 cases a year. Some even have to deal with a multiple of these numbers, such as the ED of the city of Oslo. This hospital will

---

C. Vetterli (✉)
Vetterli Roth & Partners AG, Zug, Switzerland
e-mail: christophe.vetterli@vetterlirothpartners.com

J. Hehn et al. (eds.), *Design Thinking for Software Engineering*, Progress in IS,
https://doi.org/10.1007/978-3-030-90594-1_13

have over 300,000 ED cases a year after being finalized. It can be said that the ED process is a prominent one.

The recent pandemic has underlined the massive problems in healthcare—especially in terms of data usage, resource transparency and proactive steering ability of those organizations and patient flows. It has shown that many healthcare organizations are lacking digitalized data which are gathered in a structured, standardized, and automated way and represent reality. Hence, it makes it hard for all the key process owners such as hospitals, medical centers and the linked governmental authorities to be efficient and provide the right performance. Even there are some digital "islands," the connectivity along the value chain hardly exists. The workarounds to create, e.g., "one correct truth" of capacity takes too long and is only produced with a tremendous effort from single person. In the pandemic, many staff members were gathering the number of COVID-19 patients manually running from one ward to the other to enter their gathered information into excel. By having entered this data into an printed Excel, the information was already old, since in the 60 min running from ward to ward of a bigger university hospital, the situation has already changed again. This is happening in the year 2020.

Let us walk through a different patient experience: Imaging a 25-year-old woman, Ms. Wehrle, again having some unusual pain in the belly region for 2 days, and yesterday she has had a first consultancy through a videoconference with a medical expert having asked her specific question to exclude some life-threatening source. The medical expert has received Ms. Wehrle's permission to access her health data on her mobile phone to analyze the vital data. She recommends: "Ms. Wehrle, please observe it furthermore and go to the Emergency Department if it will not get better in the next 24 hours." Ms. Wehrle has been reassured through the short exchange with the expert. However, today she is feeling worse, so she checks the capacity of the different emergency departments in the surrounding online and sees that the ED dependance a few hundred meters is nearly empty. She already has reserved her parking lot via her parking app in front of the ED dependance. Arriving there, she is being welcomed since she checked-in to the ED via mobile and provided the basic data already automatically via check-in: "Welcome Ms. Wehrle, we have the senior physician ready in examination room 3 for you." The information from the videoconference with the physician yesterday has been transmitted upfront, and the "case" with all the needed information has reached the senior physician on time to be prepared and create hypotheses about the belly pain. He has already pre-announced to the radiology department that Ms. Wehrle will be forwarded to the diagnostics directly after having informed her about his hypotheses. He expects her to stay overnight, and in the meanwhile, a pre-reserved bed would be ready. Data has been used intelligently, software was fully embedded in the patient-oriented process, and the experts could fully focus on applying their expertise to Ms. Wehrle's need.

To be able to provide a completely different experience to the patient, the organization needs to suffer from less stress in the system. Many information which formerly has been "somewhere," arriving late to the corresponding unit, has now be gathered at the right time for the right person, and the emergency case was nearly a planned case before you have entered the door with the resources and current information at hand.

The Complexity in Healthcare

The described ED case seems simple to achieve. Why is it so hard to achieve this level of seamless patient experience via human expertise and digital automatization? Why is the potential of digital solutions hardly activated, even it could provide a seamless process, would ensure patient safety and improve the efficiency of the scarce resources? Why are the service providers and clients of the healthcare processes, such as patients, relatives, physicians, nurses having such a hard time overcoming the digital and analog fragments to provide or receive the right service, with the right information at the right time from the right expert? Healthcare is well known as one of the most complex industries and hospitals especially complex organizations (Prashant Gandhi et al. 2016). Expert organizations, such as hospitals, are often still organized as expert-oriented silos. They maintain a specific structure in each silo, which in itself already provides a certain complexity. The single silos are sometimes organized efficiently but are losing efficiency as soon as interfaces are lived with the other silos. This fragmented landscape raises complexity. Some hospitals have even underlined the different expertise fields in single buildings per expertise. The patient, therefore, needs to move from one building to another to receive his multidisciplinary service. By doing so, he moves along different systems in terms of IT infrastructure, process and medical standards and has different contact points. Another specialty is the evidence-based approach that healthcare claims to follow. It takes 17 years from innovation findings to bedside appliances (Morris et al. 2011). There is this need for 100% safety before new procedures or techniques are being applied in the everyday routine. On the one hand, it seems adequate to be sure a new approach is working in, e.g., an OR procedure, however other industries would have lost their legitimization by using so long to bring innovation to the client. The regulated environment of healthcare combined with the dynamism is another characteristic. Digitalization has been put the top of the governmental agenda. E.g., the German government has launched a 3 billion 3-year program as so-called "Digital Healthcare Act" which should help hospitals to foster digitalization (German Federal Ministry of Health 2021). What is remarkable is the funding of the accompanying change management in terms of defining the requirements for the solution has been integrated into the act. This raises another specific of the healthcare industry: It is labor intensive. Thus, the numerous experts need to be integrated into any kind of change and development. The Swiss government has even declared Design Thinking as a key pillar for innovation—also for its Digitalization initiative in healthcare, "Strategy E-health Switzerland" (eHealth Switzerland 2021). The approach to define the digital aspects in the healthcare ecosystem need to be integrative, dominated by the user's and client's perspective and therefore generating a holistic view of the processes, roles, requirements for Infrastructure.

Design Thinking in Action

Healthcare should be software-intensive, however, it is not yet an integrated, holistic software environment that is patient flow oriented. The following three cases demonstrate how Design thinking has proved to be an adequate approach to enable the transformation of healthcare into a seamless digitalized environment. The right base to define requirements from a patient flow perspective was crucial. Hence, the right expert's involvement and the right data understanding which lead to the right sustainability in innovating along with the patient flow and integrating the software development perspective, was key as base. The different examples also provide practical cases in the logic of Design Thinking in the combination of requirements engineering provided by Hehn et al. (2020) as "upfront" (example 1), "infused" (example 3) and "continuous" (example 2).

Example 1: Prototyping the New ED and Outpatient Processes at a Pediatric University Hospital

A university hospital in the German-speaking Europe had the goal to integrate the pediatric internistic and the pediatric surgical clinic into one integrated ED department. The focus was to integrate and standardize processes, infrastructures, and roles. Therefore, they decided to use Design Thinking as the approach to define the multi-sided requirements. The goal was to enable a more patient-centric ED respectively outpatient flow. In a two-phased project they have defined first the future patient flow and then deviated the architectural and the software infrastructure requirements from a more detailed process flow. The Design team was staffed with physicians, nurses, architects, and IT experts.

The different prototypes were developed through several iterations, from low-resolution to high-resolution prototypes, following the well-known design cycle (see Fig. 1).

The different iterations created by the multidisciplinary and -professional team provided a very clear preview of how the future state should be. The translation of the process prototype to the IT requirements was done continuously as parallel deviation from IT staff upfront the IT development. The following figure shows that after a first phase of understanding the design space by creating the first prototypes within the design sprints II, variations were created to deviate a first preview of requirements (IT and architectural). The final definition of requirements was done in the design sprints III when the final prototypes were ready to be diffused. The overall logic is shown in the following Fig. 2.

The patient process was the reference point, also for the IT design team, which were continuously part of the process prototyping to fully understand why specific requirements were defined. This enabled the IT team members to articulate their understanding from a software perspective directly in the process prototyping

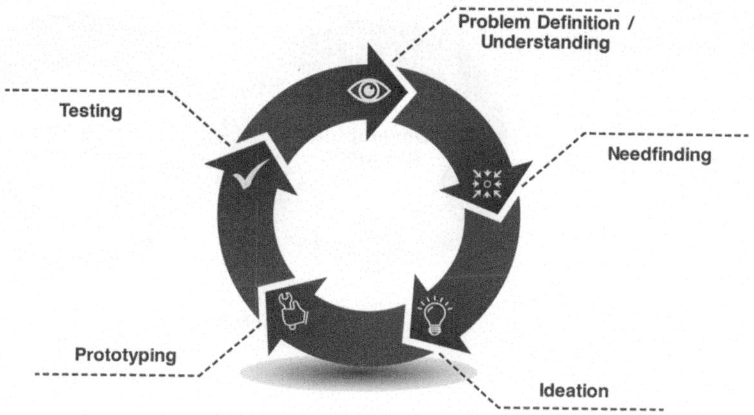

Fig. 1 Design cycle

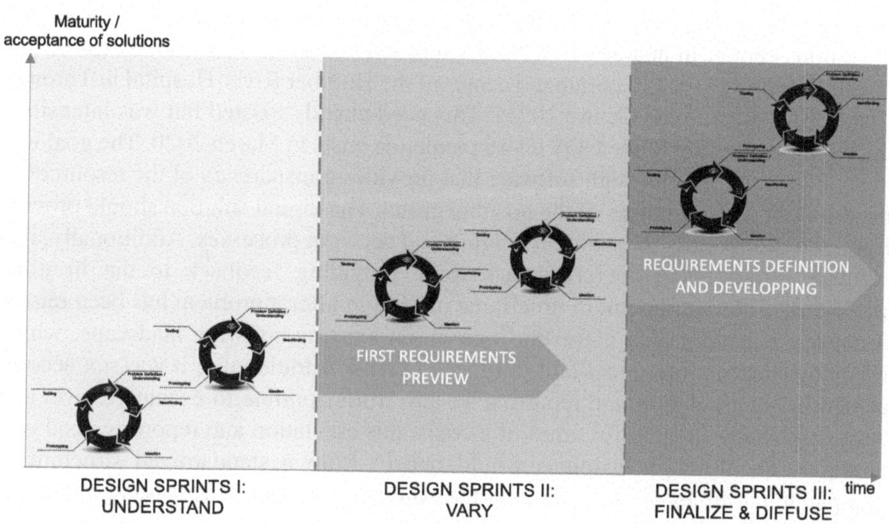

Fig. 2 Design phases to deviate requirements

sessions. Thus, this leveraged the process quality since software offered efficiency potential and assured that IT was always in line with the process perspective. The higher upfront investment from IT side paid off several times through the overall development process.

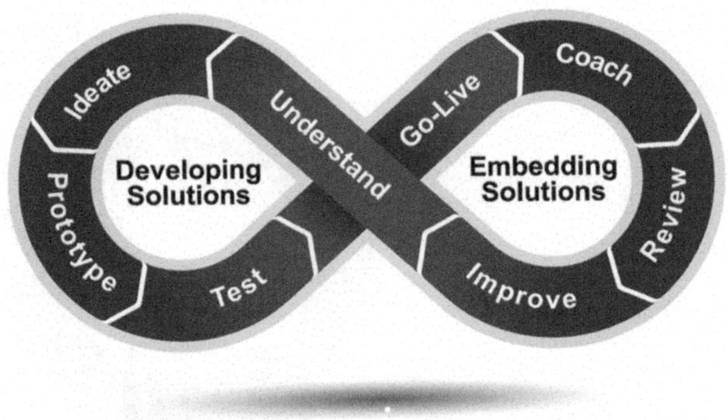

Fig. 3 Continuous developing and embedding of solutions

Example 2: Designing a Central Operations Center Software

A major hospital group in Switzerland initiated the development of a central operations center, in analogy to a NASA mission control in Houston, Texas or in a hospital setting of the GE command center of the Humber River Hospital in Toronto, Canada (GE Command Centre 2021). This need already existed but was intensified during the beginning of the COVID-19 pandemic crisis in March 2020. The goal was to centralize the data through software that provides transparency of the resources in the different core processes of the hospital group. The digital solution should provide front-end dashboards for problem-solving and decision processes. Additionally, this software should help the crisis task force providing feedback to the frontline workforce within a maximum timeframe of 90 min after a problem has been raised. It already existed a data gathering logic in an existing software landscape, which was, however not stress resistant in times of crisis. Additionally, it was not accompanied by an escalation and reporting system from frontline to executive level and back. Especially in stressful times of a crisis, this escalation and reporting need was omnipresent, although hospitals would benefit from a standardized structure in non-pandemic times as well. Finally, the software structure was lacking a connect to the overall patient flow.

A new software logic needed to be delivered and put into action fast. In parallel to a role model and standard for problem-solving and decision-making, the relevant KPIs and data flows were defined from the IT staff together with key users of the data outputs. This team could develop in parallel to the relevant roles across all functionalities. If needed, they could address questions or prototypes to the key process owners such as the OR manager, ED physician or chief nursing officer to test and feedback the software prototypes. The IT responsible observed and learned the need of the core process roles in the morning crisis meeting and used nearly every second day for a next iteration sprint. The first version of a prototype was ready after 7 days since project start—based on Microsoft's Excel but functional and directly operational. Two weeks later a first new software was developed, which was just the starting point of the development of regular next level prototype as shown in Fig. 3.

The piece-by-piece better software versions had the continuous beta-status but improved the performance of the leadership iteration by iteration. With the first levels of software pieces, the leadership meeting in the morning took about 40 min, after 4 weeks and a few iterations further, this same meeting took <15 min to address all open issues. This operations center was quickly a reference case in terms of innovation approach through design thinking, transparency of data and problem reaction time during the COVID-19 crisis (Vetterli and Roth 2020). In the second vague, the operations center was further developed and integrated more and more KPIs from elective cases since those were not present in the first vague. The Design Thinking culture of development, which was applied under extreme pressure in spring 2020, was sustainable enough to be the lead development approach also in the less stressful later stages of the crisis. It has been proved to be efficient. Hence the integrated translation from users' requirements as needs oriented requirements to a more technical, requirements engineering terminology and detail level worked seamlessly since there were back-and-forth checks on a continuous level.

Example 3: Design Thinking as Game Changer in Healthcare Digitalization

An endocrinology group of a university hospital in German-speaking Europe has observed several process restrictions by prescribing insulin to diabetes patients. First, only physicians are allowed to prescribe the medicine, which lead to process delays. Second, the prescription process was based on manual counts, which resulted in mistakes and risks. Thirdly the documentation took long and often could not be done directly after prescribing the medicine. Therefore, a spin-off of the medical university of the university hospital decided to head for a paradigm shift in core process medicine. They observed on the spot how the process was pursued and which limitations occurred. By doing so, they have started to create a new insulin prescription process based on a newly developed software.

The development process followed the core Design Thinking principles such as need orientation of patients, nurses and physicians, ideation, prototyping and testing, as well as iterating towards a better solution. The IT developers and designers focused on algorithms that analyzed the relevant diabetes items. Furthermore, they defined a prescription plan for the bedside prescription. Finally, the software developed received the CE status as a medical device. Thus, the doctor's presence in that process became obsolete and the nurses had the legitimation, together with that piece of software, to prescribe insulin to the patients. It led to a tremendous efficiency boost and raised patient safety to a maximum level. This example started to be a reference case for a paradigm shift: The development through Design Thinking core principles helped to integrate physicians in the development and create the trust that the "machine" was able to assure this life-threatening process for diabetes patients without any physician's intervention. Finally, a core hospital process was digitalized and embedded to improve patient care.

In sum, the following benefits and challenges can be derived from the three case studies:

Benefits

Healthcare is one of the only industries where the "solution" which design thinkers are working on (in general improving patient care), is merging with the client of the solution (the patient).

The needs orientation in healthcare provides the right perspective to define the precise requirements—even to substitute the physicians own performance.

The integrality of solution development and in the parallel development of precise requirements provides higher quality in less time overall.

Challenges

Bringing the right people to the table is difficult in healthcare since resources are scarce and preplanned at minimum 3 months ahead.

Unless the users have not had any prototyping experiences, it takes a special moderating effort to engage them in physical prototyping.

Too hierarchical cultures provoke that only the "highest" person solutions are being prototyped.

Summary

To sum up, the above examples were cases from the healthcare industry. Why is healthcare so relevant for the software development discussion? The fragmented environment with a tremendous amount of media and system switches, combined with a continuous change from analog to digital data and the expert organization status in a highly dynamic regulated environment, makes it a very interesting playground to apply Design Thinking as prerequisite for adequate software development processes. The extremity in so many aspects has shown that if it works in healthcare, there is a high probability that it will work in other industries as well.

The examples provide core takeaways as follows:

- Design Thinking enables adequate software development in a highly complex industry such as healthcare.

- Design thinking serves as an enabler for a common language where no translation is needed from experience prototype to software development processes.
- Parallelism in software development and the client-centric process prototyping helps foster the software development process accuracy.
- Design Thinking can be applied in extreme situations due to its characteristic speed in the application.
- Design Thinking enables to change of fundamental principles in expert organizations due to creating a trustful change through involvement in the creation process towards software development.

As shown Design Thinking applied in highly technical and software-intensive healthcare provides some explicit benefits and challenges. However, there remain some critical issues in applying design thinking. Unfortunately, the expert organization characteristics provide some limits to the design thinking application unless you engage the experts to really engage in creating tangible prototypes. An example involving a chief radiology physician who was involved in the design of a new ED department, and his radiology was not able to reduce his talking (hours long). Therefore, the prototyping time was significantly reduced, which led to prototypes that were not tested enough and did not provide enough resolution and details in the solution preview. He actively boycotted the scarce design time of the design team and afterwards disrespected the low-resolution prototypes of his colleagues. Design Thinking is strongly relying on having the people engaging in the design process. There is no shortcut and no steps that you can leave it when it comes to prototyping. Additionally, critical people who are suspicious to Design Thinking are hard to push to the first steps of prototyping together with colleagues. They like to observe the others and then comment why this is not working. Design Thinking needs the engagement of all involved parties to lead into common tested and accepted results. The observing design team members are often emotionally less attached to the innovation and skip the effort to push it all the way to innovation.

We all carry the risk to be in the patient's situation. The need of a seamless and efficient patient-centric healthcare, such as Ms. Wehrles' experience, using the potential of digitalization could be yours from 1 s to another. Design Thinking cannot solve every problem in the healthcare environment. But from a software development perspective it has proved to be a decisive enabler to foster another quality level of software development—in healthcare and certainly beyond.

References

eHealth Switzerland (2021) Strategie ehealth Schweiz. https://www.e-health-suisse.ch/politik-recht/strategische-grundlagen/strategie-ehealth-schweiz.html. Accessed 5 Feb 2021

GE Command Centre (2021) Opening of Canada's First Hospital Command Centre. https://emea.gehealthcarepartners.com/insights/17-digital-and-advanced-analytics/544-humber-river-hospital-command-centre-opening. Accessed 3 Feb 2021

German Federal Ministry of Health (2021) Digital Healthcare Act (DVG). https://www.bundesgesundheitsministerium.de/en/en.html. Accessed 5 Feb 2021

Hehli S, Gafafer T (2021) Die Schweiz hat die Digitalisierung des Gesundheitswesens verschlafen – wie sehr, zeigt ein Vergleich mit Dänemark. Neue Zürcher Zeitung. Accessed 3 Feb 2021

Hehn J, Mendez D, Uebernickel F, Brenner W, Broy M (2020) On integrating design thinking for human-centered requirements engineering. IEEE Softw 37(2):25–31

Morris ZS, Wooding S, Grant J (2011) The answer is 17 years, what is the question: understanding time lags in translational research. J R Soc Med 104(12):510–520

Prashant Gandhi P, Khanna S, Ramaswamy S (2016) Which industries are the most digital (and why)? Harvard Business Review. https://hbr.org/2016/04/a-chart-that-shows-which-industries-are-the-most-digital-and-why. Accessed 5 Feb 2021

Vetterli C, Roth R (2020) Lean operations Centre. Whitepaper Corona future management. Medizinisch Wissenschaftliche Verlagsgesellschaft, Berlin

It Takes Two to Tango: Design Thinking and Design Patterns for Better System Development

Ernestine Dickhaut, Andreas Janson, and Jan Marco Leimeister

Introduction

Smart personal assistants, such as Alexa and Siri, are becoming increasingly popular (Knote et al. 2021). They are not only used for private purposes in everyday life, but also increasingly in learning contexts (Hobert and von Wolff 2019; Winkler and Söllner 2018; Winkler et al. 2020). The use of smart personal learning assistants (SPLA) in university teaching differs from their use in private households. Both data protection regulations and the didactic added value of the system must meet the requirements of the university. Thus, many different disciplines, such as didactics, psychology, law, or computer science, are involved in the development of an SPLA for use in teaching (Hobert and von Wolff 2019). This results in multidimensional requirements for the SPLA, which must be combined into one system.

For the analysis of the complex and conflicting requirements, as well as the development of a design solution, Design Thinking offers a useful framework. Design Thinking has become a well-established approach to developing products, services, processes, and business models by improving problem solving by people from different disciplines working together to either solve a problem or develop new ideas. The approach is primarily used to solve so-called wicked problems, which are

E. Dickhaut (✉)
Information Systems, University of Kassel, Kassel, Germany
e-mail: ernestine.dickhaut@uni-kassel.de

A. Janson
Institute of Information Management, University of St. Gallen, St. Gallen, Switzerland
e-mail: andreas.janson@unisg.ch

J. M. Leimeister
Information Systems, University of Kassel, Kassel, Germany

Institute of Information Management, University of St. Gallen, St. Gallen, Switzerland
e-mail: leimeister@uni-kassel.de; janmarco.leimeister@unisg.ch

J. Hehn et al. (eds.), *Design Thinking for Software Engineering*, Progress in IS,
https://doi.org/10.1007/978-3-030-90594-1_14

considered to be particularly complex and inconsistent in finding a solution. These problems require creative and conceptual solutions, which can lead to challenges in prototypical or final product design. Thus, Design Thinking end with finding a novel solution that does not support developer to implement the discovered solution.

Design Patterns may be a valuable tool to support the implementation of those novel design solutions. By providing proven solutions for recurring problems Design Patterns are mainly established in the field of system development and support developers. We see both theoretical and practical potential in combining both approaches, namely Design Thinking and Design Patterns, to successfully develop novel artifacts in a sustainable way. The goal of our contribution is to use the advantages of the user-centered Design Thinking approach and provide a bridge to system development through the use of Design Patterns that support developers in the implementation of design solutions and is based on the following research question (RQ):

RQ: How can we combine the advantages of the user-centered Design Thinking approach with Design Patterns to design useful, novel IT artifacts?

To answer our research question, we use the example of developing a lawful SPLA. For the development of the SPLA, we use Design Thinking to extract creative solutions of possible SPLAs in an interdisciplinary team of legal experts, developers, and dialecticians. The entire study takes place in the context of a higher education lecture at a German university. We then use Design Pattern to implement these solutions in a prototype, which we then evaluate with users and legal experts using the law simulation study.

Background and Related Work

Design Thinking

Design Thinking is a structured problem-solving approach to solve real-world problems (Hehn and Uebernickel 2018). In recent years, different interpretations of design thinking have become established in both research and practice. Depending on the approach and goal, different methods and artifacts are pursued and developed. The approaches aim to improve problem-solving by having people from different disciplines work together to either solve a problem or to develop new ideas. Through novel workshop settings and usually a new environment in the workshops, the creativity of the participants is encouraged. Design Thinking focuses on the needs and motivations of the user and can be seen as a human-centered approach. In system development, we pretend too often that requirements somehow just exist and that they simply need to be elicited and documented, thus missing the true potential of fully solving the problem in a human-centric manner. Design Thinking addresses this problem and offers various tools for practical application.

Design Thinking can be seen as a process perspective, which can be used to structure the procedure from identifying the problem to finding suitable solutions.

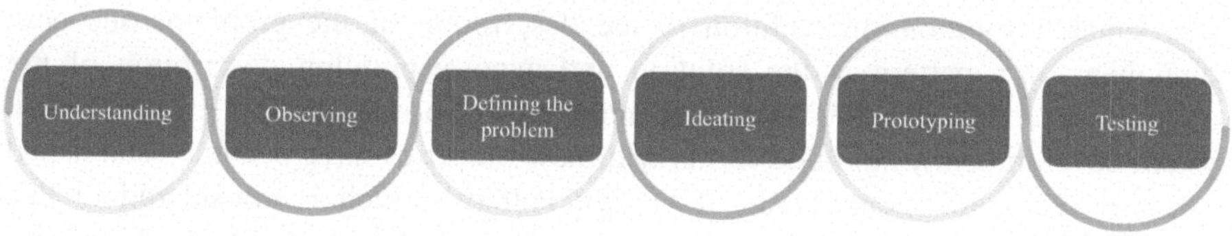

Fig. 1 Design Thinking process

Thus, many variants have become established in research and practice. In our contribution, we focus on the following phases: understanding, observing, defining the problem, ideating, prototyping, and testing (see Fig. 1).

Design Patterns

Design Patterns originally came from the field of architecture and were established in the work of Alexander (1977), in which they support architects in the designing of houses and the planning of cities. In design, architects often face recurring problems to which proven solutions can be applied. Therefore, Design Patterns codify proven solutions for recurring problems and make them usable for the future.

Thus, the use of patterns has become established in various disciplines. In Human–Computer Interaction (HCI), patterns have already been proven in many studies to teach design principles and design concepts (Borchers 2002; Compagna et al. 2007; Koukouletsos et al. 2009). In system development, patterns were first established through the Gang of Four (GoF) (Gamma et al. 1994). In addition to the previously used application areas, patterns can be used to enable a broad understanding of periphery disciplines (Wania 2019).

In system development, Design Patterns represent abstract and thus generally applicable and reusable solutions for recurring problems. Design Patterns codify "best practices" and make them usable for the future. For this purpose, a Design Pattern offers a type of "template" that follows the same structure for all patterns. In terms of content, Design Patterns do not present innovative solutions and do not reinvent the wheel but rather build on proven solutions. One Design Pattern offers solutions for many different problems. Thus, Design Patterns can counteract one of the problems identified by vom Brocke et al. (2020) in the reuse of design knowledge by codifying said knowledge into an abstract form in Design Patterns.

According to Nonaka and Takeuchi (1995), design knowledge is represented as implicit knowledge. Thus, the knowledge must be externalized to make it accessible to other people. This can be achieved either by talking about it or by writing it down and codifying it. Design Patterns externalize implicit design knowledge by codifying it and capturing it in written form. According to vom Brocke et al. (2020), Design Patterns are a component of the design knowledge base. Patterns help to find suitable solutions for existing problems. By providing information about the context of the problem, a Design Pattern helps to understand the problem in detail and thus find the most suitable solution. In addition to the description of the problem, one core component of a pattern is the solution and a corresponding description of the solution procedure. In addition to the familiar form of Design Pattern (e.g., from system development), modified forms with new purposes have become established in recent years, for example, legal Design Pattern, which is considered in this contribution.

Developing Smart Personal Learning Assistants Using Design-Thinking and Design Patterns

In the following, we demonstrate the use of Design Patterns and Design Thinking in one exemplary case, namely the development of an SPLA (Fig. 2). First, we use Design Thinking to collaborate in an interdisciplinary team consisting of dialecticians, lawyers, and developers to extract design solutions for a lawful SPLA that includes innovative didactic learning concepts and thus conveys the learning content to the user. Second, to support the development of the SPLA, we used Design Patterns. The Design Patterns were used where Design Thinking ends. We use proven, previously developed Design Patterns. The Design Patterns are developed to design AI-based assistants in a lawful way to meet strict regulation rules such as the General Data Protection Regulation (GDPR) or privacy aspects. We classify SPLA as a subgroup of AI-based assistants and thus provide developers with support for implementing creative design solutions in practice.

Finding the Design Solution Using Design Thinking

The SPLA development project starts with Design Thinking to elaborate design solutions. Thus, an interdisciplinary team from three disciplines—law, didactics, and computer science—was involved in the development of the SPLA, as these three disciplines differ in their methodologies and their viewpoint and thus in their requirements for an SPLA. To involve all disciplines equally in the development and to develop a common design solution, we follow the structured procedure of

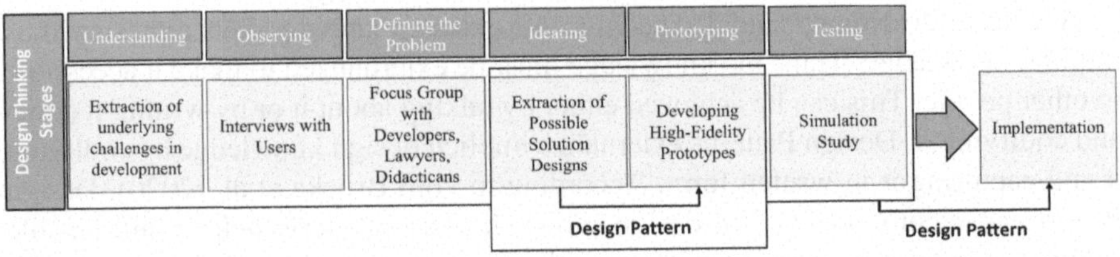

Fig. 2 Design Thinking stages

Design Thinking. In the first phase, understanding, we first conducted a joint workshop of all involved persons. The goal of the workshop was to extract the challenges in the development of a lawful SPLA.

After enabling a shared understanding of the other disciplines within the project team, the focus was placed on the user of the SPLA. For this purpose, we conducted interviews with potential SPLA users, seven students ($N = 7$). The interviews allowed us to integrate the user-centered thoughts of Design Thinking into the development. The subject of the discussion was, for example, the storage of user data. To enable adaptive learning from a didactic perspective, the user must be uniquely identifiable, and the progress must be stored and processed. But legal experts are critical of the idea, since this involves personal data that is particularly worthy of protection.

Based on the previous workshops and interviews, the project team collaborated on possible design solutions. So also, to find a solution for the storage and processing of personal data. For this purpose, the Design Pattern "deletion routine" is used during the Design Thinking process, because deletion routines present a possible solution to the problem. Many different creative solutions came out of this. As usual for Design Thinking, in this phase, solutions are also generated that cannot be implemented in practice. To avoid this, we used Design Patterns already at this stage. The Design Patterns supported the project team in thinking about the practical implementation of the design solution at an early stage as well as supporting them to implement the design solution later.

In our specific project, we created three design solutions for an SPLA in higher education. The next step was to implement one solution in an SPLA. For the transfer from the Design Thinking phase to the actual implementation, a team of three experienced developers was given both the design solution and the Design Pattern. The Design Patterns provided proven solutions for the implementation of similar design problems.

Developing the Design Solution Using Design Patterns

In the following, we describe how we developed the extracted design solutions from Design Thinking and implemented them with the support of Design Patterns in an SPLA. The used Design Patterns are a well-established approach in different projects in information system research (please see Dickhaut et al. 2020b, 2021a, b). With the Design Pattern, the previously known use in system development was expanded by providing proven solutions to legal problems in system development for AI-based assistants. The Design Patterns provide proven solutions for the development of the lawful SPLAs and were developed and evaluated in the context of another study (for more details, please see Dickhaut et al. 2020a).

The SPLA was used in the context of a university course as a complementary preparation for exams, and, together with the user, it repeated the course material in a playful way. In the learning quiz, the SPLA asked the user a question and offered

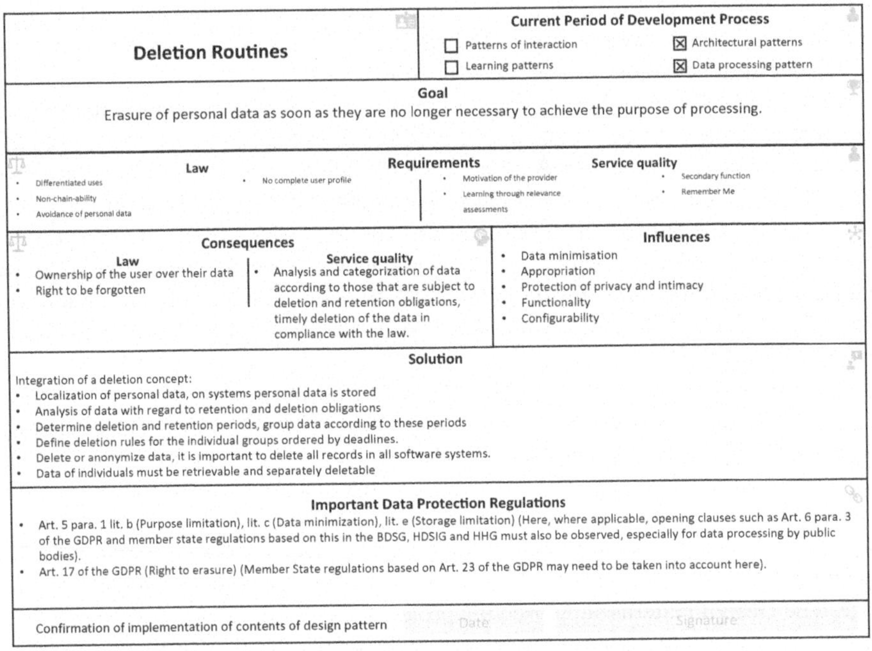

Fig. 3 Design Pattern "Deletion Routines" (Dickhaut et al. 2021a)

four possible answers, of which exactly one answer was correct. The used teaching material was based on the content of the course and was prepared together with the lecturer and other teachers of the course and implemented into the SPLA.

The Design Pattern, *deletion routine* (Fig. 3), implied that the purpose of the data processing had to be specified precisely and differentiated. It also suggested a granular consent option for this purpose. The proposed solutions from the Design Pattern were implemented in the development of the SPLA, and personalization was initially deactivated in the default settings. This allowed the user to select exactly which data storage option they wanted. For this, the user received additional information, options that explained exactly why the respective data storage was necessary, and what added value it had for achieving the goal. In addition, to ensure transparency with regard to all data processing operations, which was recommended in the Design Pattern, we designed the SPLA in such a way that users were adaptive informed about the data processing operations. The user was provided with appropriate information, not only during the first use, but whenever relevant data processing took place. To balance transparency and trouble-free use, we provided the option to select how often and how detailed they would like to be informed about data processing. In addition, as recommended in the Design Pattern, users had the option of viewing their digital self-disclosure about the storage of their data in the settings at any time. This supported the transparency of data storage.

To meet the requirements of law on purpose limitation, the Design Pattern recommended a regular check, in which personal data was still required to achieve the purpose. Deletion routines were a solution approach for this. In our SPLA, we saw the possibility of performing such automated deletion routines, especially after the end of a semester.

Overall Evaluation of the Development Process and Legal Assessment

Evaluation of the Development Process

To the best of our knowledge, with this contribution, we are the first who combine the two approaches Design Thinking and Design Pattern and demonstrate the approach with a concrete development example. In the following section, we would like to look back at the procedure and evaluate it.

Design Thinking provides a framework that enables interdisciplinary project teams to develop novel design solutions in a structured way. One of the biggest challenges at the beginning of the project was the different approaches and languages of the individual disciplines. Through the workshops in the first stages, the project team was able to broaden their own perspectives and understand the requirements of the other disciplines. This made it possible to find a common language within the project team. The structured framework supports the project team in the individual stages without limiting them to specific methods.

The combination of Design Thinking and Design Patterns lead to the development of runnable novel technologies. The challenge of combining complex and sometimes conflicting requirements in an SPLA was solved by Design Thinking. Innovative design solutions were extracted, which could then be implemented with the support of the Design Pattern. Usually, Design Thinking ends with the prototypical implementation and evaluation of a design solution. By using the Design Patterns, we build a bridge to the final implementation and support the developer in the implementation of the solution.

Legal Assessment of the Developed SPLA

We conducted a law simulation study which is a well-established evaluation method under lawyer (Rossnagel and Schuldt 2013) to evaluate the SPLA (Fig. 4). The simulation study can generally be divided into two parts. In the first part, the system was examined in a practical evaluation with users for its usability, user experience, and potential application problems. The second part of the simulation study served to evaluate the legality of the SPLA. In this second part of the study, simulated court

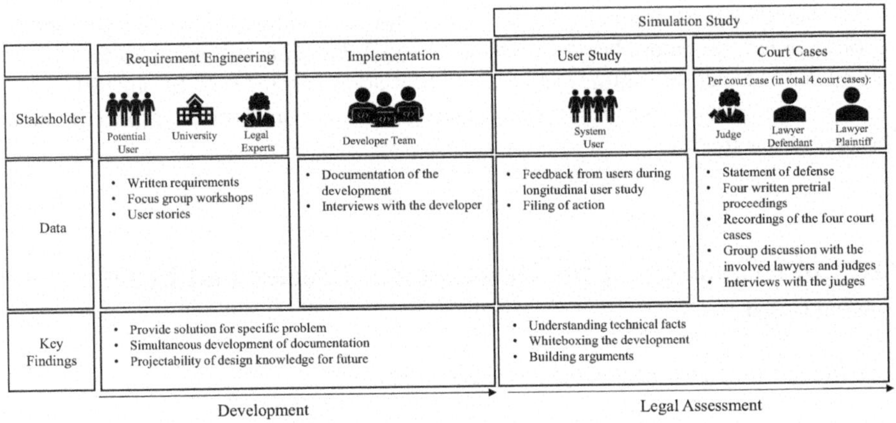

Fig. 4 Simulation study (Dickhaut et al. 2021b)

cases were conducted in which the system was evaluated with the participation of real judges and lawyers for its lawfulness.

To investigate the primary aim of the first part of the simulation study, the use of the SPLA in practice by real users was essential. For the evaluation, we offered an exam preparation course in a university basic economics course that complemented the lecture, in which students could participate voluntarily.

On three dates in a period of 2 weeks, the participants could study together with the SPLA for 1 h for the exam and repeat the acquired knowledge from the course. The offer took place on a voluntary basis. We then used questionnaires to survey student satisfaction with the SPLA. In addition, we asked in a qualitative survey for suggestions on improving the SPLA in the future and increasing its added value for students and lecturers.

Based on the user evaluation, we conducted four simulated court cases (for further information on the evaluation, please see Dickhaut et al. 2021a, b; Thies et al. 2020). The simulation study was carried out before German courts according to German and European law. Overall, six legal experts participated in our law simulation study. Among them were two judges and four lawyers that conducted the four court cases. All participants had completed the second state examination in law and already had several years of professional experience as a lawyer or judge. One participant was female, the other five were males. The oral hearings lasted 45 and 60 min. Each of the four trials involved a judge, a lawyer from the defendant, a lawyer from the plaintiff, the plaintiff, and the defendant. For the plaintiffs, we recruited voluntary participants from the first part of the simulation study to present the process as realistically as possible. In all four cases, the defendant's side was represented by the university, which used the IT artifact in the lecture course.

Before the oral hearings, written preliminary proceedings took place in which the plaintiff's lawyer set out the facts of the case and the reasons for the action and called on the defendant to refrain from using the IT artifact in university teaching. The reason for one of the four actions was the *collection of personal data beyond the purpose of processing*, as well as information about the duration and purpose of data storage. In a five-page statement of defense, the defendant's lawyer commented on the action. In the statement of defense, the lawyer referred to the *deletion routine* Design Pattern that was used in the development process of the SPLA.

Thereupon, the judge invited the participants to an oral hearing to dispute the action. In court, the judge first presented the facts of the case and discussed the reasons for the action. After the plaintiff's lawyer confirmed the facts of the case and set out the reason for the action in more detail, the two lawyers and the judge examined the facts. The reason for the action was the processing of personal data beyond the actual purpose of use, namely, the adaptation of the SPLA to the learner. The plaintiff participated in exam preparation with the SPLA last semester. The plaintiff claims to have achieved poor results in the exam preparation due to private problems. Nevertheless, she passed the exam. When the plaintiff applied for a job at the department a few months after the end of the semester and was rejected, the plaintiff saw a connection between the poor results in the exam preparation with the help of the SPLA and her not getting the job, so she claimed that the data and results of the SPLA were used beyond their intended purpose. Both parties then had the opportunity to present their side and the judge could get an impression of the situation.

According to the plaintiff, the defense lawyer came to state his case and referred in his statement to the implementation of regularly deleting routines at the end of the semester. To confirm this technically, he also referenced the Design Pattern, *deletion routine*, used in the development of the SPLA. In the course of the hearing, it turned out that the responsible chair had already deleted the data at the time of application and that it was not used beyond its intended purpose, even during its use. The negotiations ended with the pronouncement of a judgment, which was in favor of the university. In addition to the four court cases and the written correspondence, we interviewed the judges and lawyers to gain insights into the support for our pattern catalog. The interview took place directly after the end of the simulation study with all participants.

Conclusion

In this contribution, we aim to demonstrate how two approaches, Design Thinking and Design Pattern, which initially appear to be contrary, can be combined to provide comprehensive support in the development of innovative IT solutions. By first using Design Thinking for human-centered solutions, finding wicked problems, and then implementing the actual solution through Design Pattern, the support no longer ends before the actual implementation in the system but rather the developers will be further supported.

Design Thinking usually ends with the prototypical design of the design solution and therefore does not offer any support in the practical implementation. We see two possible stages in which Design Pattern can be used in Design Thinking and thus the potential of both approaches can be elaborated. First, during the analysis of the problem and the identification of possible solutions. Design Patterns show the participants the possible solution space and thus support their creativity. Second, the outcome of Design Thinking is the prototypical implementation and its evaluation of a design solution. Design Pattern provides developers beyond that with support in the practical implementation of the design solution. By using the Design Pattern in the final Design Thinking stages, we see several advantages:

- Creativity is enhanced by presenting new ideas in the Design Pattern.
- It is possible to think about which design solutions are realistic at an early stage.
- The result of Design Thinking is not *only* the design solution but also the corresponding Design Pattern used to implement the design solution.
- Law regulation rules can be considered early in the development process, thus GDPR compliance can be achieved.

The results of the evaluation also confirm this and show that the combination of both approaches leads to IT artifacts that produce good results both in user evaluation and in court. Further work should analyze the combination of Design Thinking and Design Pattern in more detail to expand the deployment scenarios of both approaches. Our contribution is intended to lay a foundation for an exciting discussion on the extension and application of Design Thinking beyond the previously known context.

References

Alexander C (1977) A pattern language: towns, buildings, construction. Oxford University Press, Oxford

Borchers J (2002) Teaching HCI design patterns: experience from two university courses. In: CHI international conference on human factors of computing systems, Minneapolis, USA, 21–25 April

vom Brocke J, Winter R, Hevner A, Maedche A (2020) Accumulation and evolution of design knowledge in design science research—a journey through time and space. J Assoc Inf Syst 23: 9–49

Compagna L, Khoury PE, Massacci F, Thomas R, Zannone N (2007) How to capture, model, and verify the knowledge of legal, security, and privacy experts. International conference on artificial intelligence and law, pp 149–153

Dickhaut E, Janson A, Leimeister JM (2020a) Codifying interdisciplinary design knowledge through patterns—the case of smart personal assistants. Design science research in information systems and technology (DESRIST)

Dickhaut E, Miedzianowski N, Jandt S, Janson A, Knote R, Leimeister JM, Roßnagel A, Söllner M, Thies LF (2020b) Handlungsbroschüre. Anforderungs- und Entwurfsmuster zur rechtsverträglichen und qualitätszentrierten Gestaltung kontextsensitiver Applikationen (AnEkA). Handlungsempfehlungen zur Gestaltung von Entwurfsmustern. Universität Kassel

Dickhaut E, Janson A, Leimeister JM (2021a) The hidden value of patterns—using design patterns to whitebox technology development in legal assessments. In: 16th international conference on Wirtschaftsinformatik (WI)

Dickhaut E, Li MM, Janson A, Leimeister JM (2021b) Developing lawful technologies—a revelatory case study on design patterns. In: Hawaii international conference on system sciences (HICSS), vol 54

Gamma E, Helm R, Johnson R, Vlissides J (1994) Design patterns: elements of reusable object oriented software. Addison-Wesley Professional, Boston

Hehn J, Uebernickel F (2018) Towards an understanding of the role of design thinking for requirements elicitation-findings from a multiple-case study. In: Americas conference on information systems (AMCIS)

Hobert S, von Wolff RM (2019) Say hello to your new automated tutor–a structured literature review on ppedagogical conversational agents. International conference on information systems (ICIS)

Knote R, Janson A, Söllner M, Leimeister JM (2021) Value co-creation in smart services: a functional affordances perspective on smart personal assistants. J Assoc Inform Syst 22(2)

Koukouletsos K, Khazaei B, Dearden A, Ozcan M (2009) Teaching usability principles with patterns and guidelines. In: Kotzé P, Wong W, Jorge J, Dix A, Silva PA (eds) Creativity and HCI: from experience to design in education, IFIP—international federation for information processing, vol 289. Springer, Boston, pp 159–174

Nonaka I, Takeuchi H (1995) The knowledge-creating company. How Japanese companies create the dynamics of innovation. Oxford University Press, Oxford

Rossnagel A, Schuldt M (2013) The simulation study as a method of evaluating socially acceptable technology design, pp 108–116

Thies LF, Dickhaut E, Janson A, Roßnagel A, Leimeister JM, Söllner M (2020) Die Simulationsstudie als Evaluationsmethode. Interdisziplinäre Evaluation eines smarten persönlichen Assistenten. Datenschutz und Datensicherheit (DuD)

Wania C (2019) Exploring design patterns as evaluation tools in human computer interaction education. MWAIS 2019 Proc 9

Winkler R, Hobert S, Salovaara A, Söllner M, Leimeister JM (2020) Sara, the lecturer: improving learning in online education with a scaffolding-based conversational agent. In: Bernhaupt R, Mueller FF, Verweij D, Andres J, McGrenere J, Cockburn A, Avellino I, Goguey A, Bjørn P, Zhao S et al (eds) Proceedings of the 2020 CHI conference on human factors in computing systems. ACM, New York, pp 1–14

Winkler R, Söllner M (2018) Unleashing the potential of chatbots in education: a state-of-the-art analysis. In: Academy of management proceedings

Epilog: From Requirements Engineering to Design Thinking

Manfred Broy and Walter Brenner

Epilog

Design Thinking has been initially created as a method offering cognitive, strategical, and practical processes to develop design concepts for products, buildings, machines, and communications. Initially, Design Thinking was created with emphasis on the physical world. However, nowadays, systems and system designs, including services and all kinds of complex, distributed cyber-physical systems, gain their creative power and their innovation to a large extent from the abilities and possibilities of advanced software, including dedicated user interfaces, data analytics, and service engineering also based on concepts of artificial intelligence. As a result, the amount and possibilities of innovation are determined to a large extend by software and by the close interaction between software and the physical world.

Developing and designing software and software-intensive systems is significantly different from the creation of mechanical and physical products and processes. The software permits a much larger space of solutions, the software does not need a production process, and it can be changed more flexibly—even after deployment. These differences are of high significance, in particular, for the goals of approaches like design thinking.

Traditionally and historically, at the beginning of the application of software, mainly solutions for existing analog processes were implemented by software. Consequently, software was not so much innovative but rather an improvement with respect to precision, efficiency, and the ability to handle large structures. Only

M. Broy
Department of Informatics, Technical University of Munich, Munich, Germany
e-mail: broy@in.tum.de

W. Brenner (✉)
Institute of Information Management, University of St. Gallen, St. Gallen, Switzerland
e-mail: walter.brenner@unisg.ch

J. Hehn et al. (eds.), *Design Thinking for Software Engineering*, Progress in IS,
https://doi.org/10.1007/978-3-030-90594-1_15

incidentally, it turned out that the power of software systems brings in a completely new impact on innovation, and creative solutions. The reasons for the remarkable power of software are manifold. They include the ability to relate and integrate data and functionality from other applications. Much more detailed ways to analyze data are a further reason. And finally, the automation of processes on the one hand and on the other hand the potentiality to create powerful user interfaces is perhaps most significant.

This historical development of software already is an indication for a kind of discrepancy between the conventional development of software, which rebuilds solutions in the physical world by software, and creative (development of) software, which allows for completely new applications based on the power of nowadays communication networks, cyber-physical applications, and inter-operability between very different systems. This allows solutions and approaches very different from what can be achieved in a world without software.

This difference is also expressed in the methodological approaches to software development. Conventional systems are developed in a rather obvious way: Capturing requirements by studying existing physical solutions and translating them into requirements for software solutions. In this "old" style of building software, there is not so much space for innovation. Known solutions and processes are transferred to software, making them automatic, more efficient, and more precise.

This form of development and proceeding is in contrast to a more creativity-oriented development style. Going in this direction is very challenging. Development teams have, on the one hand, to understand the rich possibilities and opportunities of digital technology and, on the other hand, to relate those possibilities to the ultimate demands with respect to the problem and application area under consideration, and understanding the needs of the customer. In particular, dealing with wicked problems and—as a result—tricky solutions might bring in completely new aspects and possibilities as results of a careful combination of the power of software technology and digitization with deeper insights into the needs of the user community with respect to the particular problem domain.

These two completely different ways to approach the development of software are, in particular, visible in the phase of requirements engineering. In a conventional style, requirements engineering means just to study the physical approaches that should be reflected in a digital solution: In the requirements. This is a demanding step. Nevertheless, because formal approaches are based on human perception and judgment, they have been deeply analyzed and reflected in the requirement artifacts of systems. A typical example could be found in health systems: If a doctor communicates with a patient about his symptoms to come up with a hypothesis about the particular disease, it is a very difficult step to try to replace this task—at least partially—with software which does the analysis. So difficult it may be, it is just an attempt to replace a process of human investigation with some software. This leads to the classical, conventional field of requirements engineering, where this formalization is the challenge.

However, by such a proceeding, the critical question is whether we are able to find out about the possibilities of digital solutions to come up with completely new ways of creating systems and solving problems. This is much closer to the goals of design thinking.

Therefore, we believe that it is an important step to find out about a kind of synthesis between design thinking and requirements engineering. It is less simple than it looks at the first moment. Design thinking, originally, was not specifically aiming at software-intensive systems. Therefore, it is necessary to analyze a lot of interesting aspects of the specific techniques of software development and to see how this can be combined with questions of design thinking.

It covers, in particular, aspects such as the typical artifacts in requirements engineering for software systems, questions of development processes for software such as Scrum, the role and potentials of data analytics, the particular roles for requirements engineering, for design thinking, and user experience, and how these fit together, but also special issues such as change management, strategic road maps, managing tensions as well as special application areas. These include shop floor support systems, digital platform creation, healthcare, and smart personal assistant.

The contributions span a wide spectrum, however, being far from completely covering the subject. Nevertheless, they show in a quite impressive way the many topics to consider.

This is what this volume is about. It contains an interesting number of contributions along these lines. It covers a broad spectrum of questions of the application of design thinking to requirements engineering and vice versa, including questions of processes, development, organizational structures, and special issues in the development such as treating conflicts or defining roles, for example.

A very promising and interesting contribution is asking for a new kind of discipline of what is called “digital design,” a field which is already practiced today to a large extent, but there is still a lot to investigate. Certainly, for instance, a key result is to find an approach being a good synthesis between classical requirements engineering and design thinking. This objective also expresses explicitly the general goal of this volume: To bring together two important techniques into a discipline that is about to change the world: digitization and its major discipline for creating digital systems: digital design thinking.

The manufacturer's authorised representative in the EU is Springer Nature Customer Service Centre GmbH, Europaplatz 3, 69115 Heidelberg, Germany. If you have any concerns regarding our products, please contact ProductSafety@springernature.com

Printed and bound by CPI Group (UK) Ltd, Croydon, CR0 4YY
22/09/2026
02220982-0007